In Deeper Waters

Pisces V on the launch platform (depth 15 m, photo Kerby).

*E. H. Chave and
Alexander Malahoff*

In Deeper Waters

PHOTOGRAPHIC STUDIES

OF HAWAIIAN

DEEP-SEA HABITATS

AND LIFE-FORMS

UNIVERSITY OF HAWAIʻI PRESS

HONOLULU

A publication of the University of Hawai'i Press
pursuant to National Oceanic and Atmospheric
Administration Award Nos. NA66RUO186 and
NA46RUO145

© 1998 University of Hawai'i Press
All rights reserved
Printed in Canada by Friesens
03 02 01 00 99 98   5 4 3 2 1

Library of Congress Cataloging-in-Publication Data
Chave, E. H.
    In deeper waters : photographic studies of Hawaiian deep-sea habitats and life-forms / E. H. Chave and Alexander Malahoff.
       p.   cm.
    Includes bibliographical references and index.
    ISBN 0-8248-1946-2 (cloth : alk. paper). —
    ISBN 0-8248-2003-7 (pbk. : alk. paper)
    1. Deep-sea animals—Hawaiian Ridge.
    2. Benthic animals—Hawaiian Ridge.
    3. Geology—Hawaiian Ridge.
I. Malahoff, Alexander, 1939–   II. Title
QL125.5C48   1998
591.77'49—DC21   97-46936
                                                CIP

University of Hawai'i Press books are printed
on acid-free paper and meet the guidelines for
permanence and durability of the Council
on Library Resources

Design by Barbara Pope Book Design

Contents

	ACKNOWLEDGMENTS	VI
	INTRODUCTION	1
1	DEEP-SEA GEOLOGY	9
2	DEEP-SEA ECOLOGY	23
3	DEEP-SEA ANIMALS	28

Sponges (Porifera) 28
Corals and Related Species (Cnidaria) 30
Sea Stars and Related Species (Echinodermata) 43
Crustaceans (Arthropoda) 53
Sea Snails and Related Species (Mollusca) 56
Fishes (Chordata) 58

TABLE 1	77
A Taxonomic List of Animals . . .	
REFERENCES	111
INDEX	123

Acknowledgments

WE ACKNOWLEDGE THE ASSISTANCE of the following experts who aided in the identification of specimens or photographs:

bryozoans: A. Cheetham

cnidarians: C. Arneson, T. Bayer, S. Cairns, D. (Dunn) Fautin, S. France, R. Grigg

crustaceans: K. Baba, F. Chace, J. Haig, R. Manning, R. Moffitt, W. Newman, A. Williams

echinoderms: B. Burch, D. Devaney, L. Eldredge, G. Hendler, C. Messing, R. Mooi, D. Pawson, M. Roux, C. Young

fishes: W. Anderson, B. Mundy, D. Cohen, T. Iwamoto, G. Johnson, S. Ralston, J. Randall, R. Rosenblatt, M. Seki, P. Struhsaker, K. Sulak, A. Suzumoto, K. Tighe

mollusks: E. Hochberg, E. A. Kay, C. Roper, R. Young

sponges: H. Reiswig, K. Tabachnick

worms: J. Brock

We also wish to acknowledge the principal investigators on whose missions some of the photos were taken: C. Agegian, P. Colin, J. Cowan, S. Dollar, S. France, M. Garcia, R. Gooding, R. Grigg, R. Holcomb, D. Karl, P. Lobel, F. Mackenzie, J. Maragos, G. McMurtry, J. Moore, L. Mullineaux, J. Polovina, S. Ralston, T. Russo, P. Scheuer, J. Wiltshire, and C. Young. Many thanks to the Hawai'i Undersea Research Laboratory (HURL) submersible pilots T. Kerby and the late B. Bartko for technical support; to the HURL data team of J. Culp, L. Hu, and B. Muffler for help with photographs, videoprints, and book layout; and to P. Verlaan for final manuscript comments.

Introduction

THE VAST DEEP-SEA AREA called the Hawaiian Ridge is a dark, cold world of undersea mountains, valleys, and sediment plains and such unique geological features as pillow lava flows and volcanic vent fields. It is a sparsely populated world, home to unusual animals that do not migrate upward into sunlit waters (fig. 1). To date submersibles have explored only a tiny portion of the ridge's 2,600 kilometers, but several important discoveries have already been made of previously unknown geological processes and animal communities, and surely equally exciting finds await future explorations.

Scientists first studied the deep waters around the Hawaiian Islands in the early 1900s, lowering dredges, corers, trawls, and traps from ships to collect rocks, sediments, and animals. The development of scuba equipment in the 1940s allowed the direct observation of marine environments to depths of about 30 meters. In Hawai'i, these methods of investigation were augmented in 1965 when the submersibles *Asherah* and *Naia* briefly visited the Islands and mapped the submarine terraces and escarpments off western O'ahu to a depth of 200 meters. It was during these dives that scientists observed for the first time some of the many species of shallow-water Hawaiian fishes living below scuba diving depths. In the 1970s, the submersible *Star II* operated to depths of 400 meters around the high Hawaiian Islands. While its primary mission was to investigate a large deep-sea coral bed off Makapu'u, O'ahu, and harvest precious corals (Grigg 1977, 1988), the scientists aboard studied other deep-sea animal groups as well (Clarke 1972; Grigg and Bayer 1976).

When the National Oceanic and Atmospheric Agency's Undersea Research Program and the University of Hawai'i established the Hawai'i Undersea Research Laboratory (HURL) in 1980, *Star II* was redesigned. Renamed *Makali'i* (fig. 2), it dove in waters off Enewetak

Opposite:
Figure 1. A sea toad *Chaunax umbrinus* and toadstool coral *Anthomastus fisheri* on an otherwise barren lava talus slope (fish 15 cm, depth 333 m, photo 232-36 McMurtry).

Figure 2. Divers loading equipment on the submersible *Makali'i* (photo Dollar).

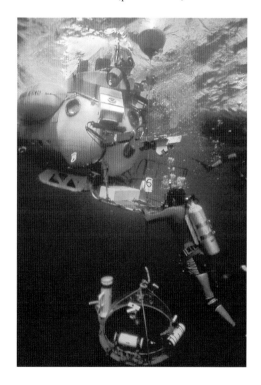

Atoll, the Hawaiian Archipelago, and Johnston Atoll. Soon, however, scientists needed to explore depths greater than 400 meters. So, in 1986, the submersible *Pisces V* (fig. 3) was purchased and redesigned, and it has been diving to depths of 2,000 meters since 1987. In addition, deeper-diving American and Russian submersibles such as *Alvin, Sea Cliff,* and *Mir* have visited the Islands and explored Hawaiian ocean environments to depths of 6,000 meters.

Figure 3. *Pisces V* (photo Kerby).

In order to dive in HURL submersibles, scientists must first submit proposals describing their intended deep-sea research program. Once a proposal is accepted, a location for the dive is chosen that will best accomplish the research objectives. Given current research interests, dive sites have tended to cluster in certain areas, leaving others—for example, the areas off Kaua'i and the Northwestern Hawaiian Islands—still relatively unexplored. (See the map of the southern Hawaiian Archipelago in fig. 4, where the dive sites of *Makali'i* and *Pisces V* are represented by red dots.) On HURL dives scientists have studied such phenomena as underwater landslides and volcanic eruptions, ancient coral reefs, and fisheries as well as the effects of geochemical cycles and marine pollution.

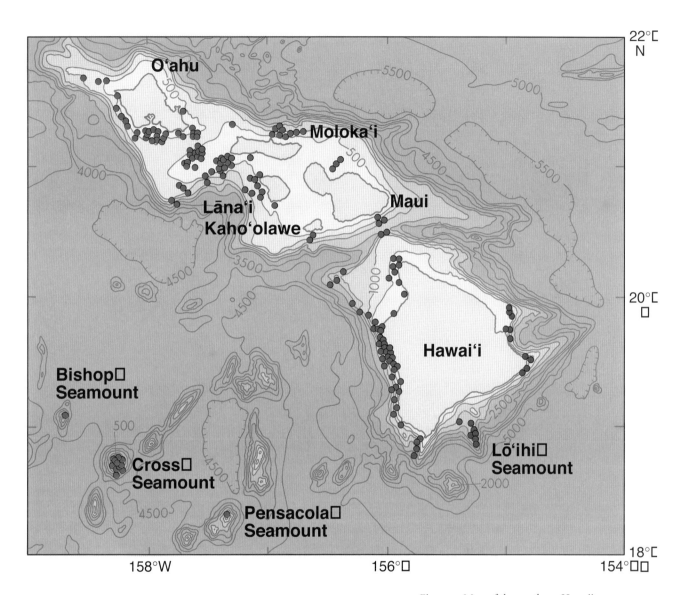

Figure 4. Map of the southern Hawaiian Archipelago showing submersible dive sites as red dots (depth contours in meters).

Figure 5. *Pisces V* being launched from the research vessel *Kaʻimikai-o-Kanaloa* (photo Malahoff).

Diving in HURL Submersibles

The command capsules of HURL's two submersibles, *Makaliʻi* and *Pisces V,* are steel spheres with cone-shaped Plexiglas ports set into the hull. Pilots and observers enter the submersible at the surface and remain sealed in the sphere at 1 atmosphere (sea level) pressure for the duration of the dive. As the submersible descends to the ocean floor, water pressure on the hull increases steadily (1 atmosphere every 10 meters). A forceful demonstration of the effects of this increasing pressure is often made by submersible pilots, who place foam cups in the submersible's collection basket and leave them there for the duration of the dive. Since the basket is outside the sphere and therefore unprotected, water pressure compresses the cups to a mass of material the size of a thimble.

Until recently, a ship towed HURL's submersibles to dive sites on a catamaran dubbed the LRT (short for launch, retrieve, transport). On site, the LRT carries the submersible down to a depth of 20 meters, thus avoiding possible damage to its hull, instruments, and samples resulting from surface wave action (fig. 2). Lowering and raising the LRT through the water are controlled by a scuba-diver pilot from the catamaran's cockpit. When the pilot releases air from ballast tanks, it is replaced by seawater, allowing the LRT to descend. When seawater is replaced by compressed air pumped from storage tanks, it ascends.

The LRT is an ideal submersible platform for use around the Hawaiian Islands. However, if the submersibles are towed long distances, equipment is often damaged by rough seas, and the operations crew is forced to make major repairs. In order to facilitate more effective research ventures at greater distances, a ship was purchased and redesigned to house a submersible on deck and to deploy it from the deck directly into the ocean by means of an articulated A-frame. *Pisces V* now operates from the 70-meter-long research vessel *Kaʻimikai-o-Kanaloa* (fig. 5).

Most submersibles are propelled through the water—forward, backward, right, or left—by electrically powered thrusters. As with the LRT, air blown from the ballast tanks allows the submersible to sink, and compressed air pumped back into the ballast tanks allows it to rise. On the deepest dives, iron ballast pellets provide additional weight during the descent. Once the bottom is reached, some of the pellets are discarded, allowing the submersible to hover above the sea floor. At the end of the dive, more pellets are discarded, enabling a more rapid ascent. This last discharge is especially necessary when the vehicle is carrying a load of samples.

The ballast tanks and the batteries that run the thrusters, lights, cameras, and mechanical arms are attached to a framework covered by fiberglass panels (the yellow coverings in figs. 2, 3, and 5). These covers are highly visible and allow the ship's crew to spot the submersibles more easily when they reach the surface. Insulated wires penetrate the hull that allow the submersible pilot and the observers to manipulate the external mechanical arms and cameras from inside the capsule.

Air inside the submersible's command sphere is circulated through scrubbers that remove excess carbon dioxide. Oxygen, housed in tanks, is injected into the capsule's atmosphere, replacing the oxygen consumed in breathing. Enough oxygen is stored in these tanks for three people to breathe for about 70 hours. Since most dives last only six or eight hours, the oxygen supply is more than sufficient.

The command sphere of *Makaliʻi* measures 5 feet in diameter and is large enough to hold the pilot and one observer. The command sphere of *Pisces V* is 7 feet in diameter and holds the pilot and two observers. Although the working space is cramped, only one person has ever reported feeling claustrophobic. The crew is extremely busy throughout the dive, constantly gathering information—observing the sea floor or operating cameras, instruments, manipulators, and other specialized scientific equipment.

In order to ensure a safe dive and avoid being trapped on the ocean floor, there are two types of situations that the pilot must avoid at all costs. The first is venturing into an undersea cave and losing power, which is highly unlikely because pilots will not take the submersibles under overhangs or into caves. The second is getting tangled in fishing lines dangling from abandoned traps or cables resting on the sea floor, and pilots must be constantly on the lookout for these dangers. In fact, in areas where such problems are known to occur, pilots will often review videotapes of previous dives and map the location of lines and cables. Should a submersible get into trouble, rescue operations would be mobilized immediately. A second submersible or a remote-controlled vehicle can be flown to the accident site, dive, and rescue the trapped vehicle all within 70 hours.

When operating in Hawaiian waters at depths of less than 300 meters, the submersible's sphere grows increasingly hot and humid. Not only is the water temperature outside about 15°C–28°C, but the operating equipment also generates heat, and heat cannot be vented until the dive is over. Conversely, at depths of over 400 meters, the water temperature outside the hull grows increasingly colder (dropping, for example, to 4°C at 1,000 meters and about 2°C at 2,000 meters). As it gets colder and colder inside the hull, heavy clothing must be worn.

Because deep-sea dives in submersibles are very expensive, locating dive sites from the surface saves money as well as time. In order to locate dive sites, HURL first uses the mapping system Sea Beam. An array mounted to the underhull of the support ship emits sound waves, which are reflected off the ocean floor and received back on the support ship. The information so obtained allows the construction of high-resolution bathymetric maps and images of the ocean bottom. The results are then correlated with photos taken by HURL's "bottom camera," which is tethered to the support ship and takes about thirty-

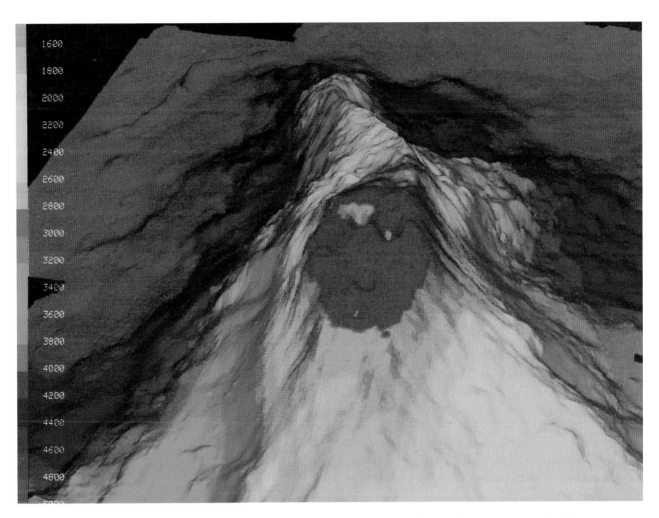

Figure 6. Sea Beam image of Lōʻihi showing the summit, craters, and slopes (Malahoff and Smith).

two hundred photographs per lowering, and dive sites selected. Figure 6 reproduces a map constructed in this way, showing the craters and other features of the Lōʻihi submarine volcano (Fornari et al. 1988).

As the submersible descends, daylight fades to twilight at about 300 meters and then to darkness. In order to conserve battery power, the running lights are turned on when the submersible reaches the bottom; the bright photo lights are reserved for use only during photo sessions. It should be noted, however, that, even if the water is extremely clear, the brightest submersible lights can illuminate objects only within a range of 4 meters from the camera. (This is why, in the deep ocean, seamounts and other features cannot be photographed from a distance and mapping techniques such as Sun Beam must be utilized and paired with the close-up bottom photographs.) Cloudy water, small or distant subjects, and poor illumination yield poor-quality images. As a consequence, fewer than 1 percent of the over thirty thousand slides in the HURL archive are of sufficiently good quality to be published. Most of the good images obtained are of objects larger than a baseball and within 2 meters of the camera. And, to get such images, the crew must have the time to turn on the lights and maneuver within focusing range.

Determinations of animal taxa observed from submersibles and comparisons to photographic, videotape, and specimen identification have been offered by several authors (Macurda and Meyer 1976; Ohta 1983; Ralston et al. 1986; Chave and Jones 1991; Chave and Mundy 1994; Roux 1994). Rock and animal specimens collected by the HURL submersibles have been identified by the many experts we acknowledge at the beginning of this book. But because many photographs and videotapes show animals that were not collected, even when image quality is good, their identification must remain tentative, based on the educated guesses of our experts.

1 *Deep-Sea Geology*

MAGMA FROM THE AESTHENOSPHERE, a hot plastic layer lying about 60 kilometers beneath the earth's surface, or lithosphere, builds volcanic islands. The lithosphere itself is composed of rigid plates. Oceanic ridges such as the East Pacific Rise and trenches such as the Mariana Trench form the boundaries of the plates. Most of the active volcanos in the world occur either in zones of fissuring (the trenches) or along collision boundaries (the ridges) between plates. However, some volcanos, such as those that are building the Hawaiian Ridge, are located in the middle of plates. The volcanos on the island of Hawai'i and Lō'ihi submarine volcano are examples of such surface and submarine midplate "hot spot" volcanos.

All mid–Pacific Plate islands began as submarine volcanos erupting from fissures on the ocean floor. This plate is currently transporting the Hawaiian island chain toward the northwest, away from a fixed hot spot in the aesthenosphere currently located under the island of Hawai'i. Geologists have clocked the movement of the Pacific Plate at a rate of 1 meter per 100 years (Jackson et al. 1972).

Today the oldest seamounts, islands, and atolls are in the northwest part of the Hawaiian island chain, farthest away from the hot spot (Macdonald et al. 1986). The active, emerging volcano Lō'ihi is currently closest to it. Midway, which formed 28 million years ago over the site now occupied by Lō'ihi, is 2,600 kilometers northwest of the hot spot.

Deep water separates the individual islands along the chain because the hot-spot magma supply has waxed and waned as the plate has moved over it. As the Hawaiian island volcanos moved farther to the northwest, they lost their magma source and became extinct (Eaton 1962; Clague and Dalrymple 1987). Cross and other seamounts to the southwest of the Hawaiian volcanos were formed elsewhere, built on or near the East Pacific Rise about 98 million years ago, briefly reaching the surface during formation and then subsiding (Malahoff et al. 1985).

Coral reefs form around tropical and subtropical volcanic islands after they emerge from the sea (fig. 7). Hawaiian volcanos sink (about 2.6 millimeters per year) because of the heavy load they impose on the lithosphere (Moore and Clague 1992), and the coral reefs surrounding them erode over time and sink as the islands subside. Johnston Atoll and islands or atolls such as Kure, Midway, and French Frigate Shoals are complex limestone structures capping sunken volcanos that arose from the sea and eroded more than 20 million years ago (Stearns 1966). Fringing coral reefs partially ring the high volcanic islands of Kaua'i, O'ahu, Maui, Moloka'i, Lāna'i, and Hawai'i. Extensive barrier reefs and lagoons are absent from the high Hawaiian Islands (James

Figure 7. Living patch reef (depth 10 m, photo Chave).

1983). Fringing reefs occur in a series of steps down from the shoreline to a depth of 1,800 meters, living reefs to 100 meters, and dead reefs below 100 meters (Moore and Campbell 1987).

When molten lava extrudes onto the ocean floor from fissures, the surface of the flow cools instantly in the cold seawater, and only the interior remains molten. The form that these submarine lava flows take may be rope-like (fig. 8), blocky (fig. 9), columnar (fig. 10), or tubular (figs. 11 and 12). Frequently, lava tubes containing molten lava break, and the magma flows out to solidify into rounded pillows (fig. 13). One pillow containing lava may break into another pillow and then another as the molten lava flows down the slope. Extensive fields of pillows are built in this manner. Pillows vary in size from 1 to 3 meters in diameter, whereas the other forms of lava range in size from a few centimeters to several meters.

Figure 8, *top left:* Ropy lava with solitary stony corals (depth 336 m, corals 0.5 cm, photo 114-34 McMurtry).

Figure 9, *bottom left:* Blocky lava and moray eel (depth 202 m, eel's head 16 cm, photo 292-46 Moore).

Figure 10, *top right:* Columnar basalt (depth 1,430 m, photo 5288-12 Holcomb).

Figure 11, *bottom right:* Lava tubes (depth 2,000 m, photo 5079-10 Malahoff).

Figure 12. Elephant's trunk lava tube (depth 202 m, photo 292-49 Moore).

Figure 14. Ancient coral reef (depth 750 m, photo 5068-40 Moore).

Figure 13. Pillows (depth 400 m, photo 113-77 McMurtry).

Figure 15. Limestone bench (depth 245 m, photo 149-62 Chave).

Figure 16. Rippled sand at base of limestone bench (depth 343 m, photo 208-18 Eldredge).

Figure 17. Grooved limestone (depth 355 m, photo 195-43 Ralston).

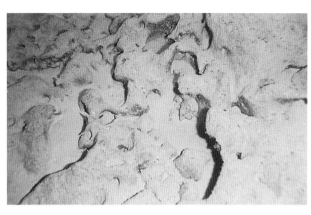

Figure 18. Holes and pockmarks in limestone, scorpionfish *Setarches guentheri*, and ornate sea star *Mediaster ornatus* (fish 25 cm, depth 400 m, photo 5066-10 Moore).

Such geological features as ancient coral reefs (fig. 14), wave-cut limestone benches (fig. 15), and beach rock form above or just below sea level and can be found off all the islands to depths of 2,000 meters. Also, both the erosion and deposition of minerals on rock surfaces are very common in the deep sea (James 1983).

Because of coastal and submarine erosion, a mixture of sediment types covers Hawaiian island submarine slopes, the channels between the islands, and the abyssal apron around them. The two major components of these submarine sediments are dark sand and clay, derived from the fracturing and chemical erosion of basaltic rocks; and white limestone sand and silt, derived from coral reef erosion and the skeletons of the many small animals that live on the ocean floor or in the water column. If the local currents are weak, plains of fine sediment are found in both shallow and deep areas of the Hawaiian Ridge. Strong to moderate currents produce fields of rippled and winnowed sand or gravel (fig. 16). Very strong currents scour the bottom clean of sediment, especially in the narrow channels between the high Hawaiian Islands.

Sand sliding down submarine limestone cliffs is abrasive and forms grooves or flutes in the rock (fig. 17). Older limestone becomes pitted or pockmarked, eroded by the chemicals in seawater, and contains holes of various sizes (fig. 18). Talus composed of broken reef material or basalt fragments covers many areas below steep cliffs (fig. 19). If conditions are right, calcium carbonate dissolved in seawater precipi-

Figure 19. Basalt and limestone talus (depth 245 m, photo 225-88 Karl).

tates as limestone, forming crusts on the bottom (fig. 20). Limestone also cements talus, sand, or ash fragments together to form hardpans (fig. 21). Other minerals encrust existing rocks under proper conditions. For example, dissolved manganese may come out of solution and form shiny, black, botryoidal crusts that overlie hard substrates (fig. 22).

Geological processes unique to the volcanos, seamounts, and islands visited by the submersibles are described in the following sections.

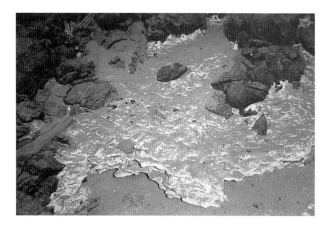

Figure 20. Limestone crusts on basalt (depth 1,270 m, photo 5065-37 Moore).

Figure 21. Cemented ash and talus (depth 755 m, photo 5076-55 Moore).

Figure 22. Manganese crust, primnoid gorgonian, and brittle stars (primnoid 40 cm, depth 970 m, photo 5234-16 Bertram).

The Island of Hawai‘i

The island of Hawai‘i has the largest land mass of the central Pacific Islands. Its volcanos reach heights of over 3,400 meters above sea level and slope beneath the sea to a depth of 6,000 meters. Over time, the island has undergone considerable change. It sank rapidly during the period of peak volcanism in the Pleistocene. Coral reefs that ringed parts of the island drowned, some sinking to depths of over 1,500 meters (Szabo et al. 1991; Jones 1993). Volcanic eruptions and earthquakes have caused slumping to the extent that Hawai‘i's undersea talus-covered slopes are five times larger than the total land area of its islands. Perhaps the largest talus and debris landslide in the world is off Hawai‘i's Kona District (Moore et al. 1989).

Geologists in the submersible have studied submarine lava flows, canyons, ledges, and aquifers off the eastern slopes of the island of Hawai‘i. The first such area visited was to the southeast, where vast areas of undulating terrain formed by different types of submarine lava flows along the Southeast Rift of Kīlauea volcano were found (McMurtry et al. 1983). The older flows to the north are smoother than those off Cape Kumukahi at the southeastern point of the island. The form of the flows varies from ropy lavas and intact pillows and tubes to broken talus, beach gravel, and sand. Blocky lava that flowed into the sea during a rift zone eruption in 1960 covers the Cape Kumukahi offshore area (Malahoff and McCoy 1967; Fornari et al. 1978).

The second area visited was the Mauna Kea Ledge, where ancient lava flows from Mauna Kea volcano have formed a wide, 300-meter-deep shelf off the eastern part of the island. The submarine part of the Mauna Kea Ledge is now a flat plain covered by sediment and scattered limestone outcrops (Gooding et al. 1988). Limestone formed from fine sediment and lithified by calcium carbonate deposits covered the lava flows and eroded to form smooth mounds (fig. 23). Seaward, the ledge drops off into Puna Canyon, where the slope consists mostly of exposed layers of basalt (Wiltshire 1982).

The submersibles traversed the western and southern slopes of Kohala, Hualālai, and Mauna Loa volcanos. Scientists determined island submergence rates of 2.6 millimeters per year and characterized the lava flows of the three volcanos. Off this part of the island, lava flows and coral reefs of different ages cover most of the area from the shoreline to depths of 50 meters. Deeper than 50 meters, lava flows, talus slopes, and limestone terraces are interspersed with sandy slopes and fine sediment terraces (Moore and Clague 1992). Along the west-

Figure 23. Black coral *Antipathes* sp. 1 and several cidarid urchins *Stereocidaris hawaiiensis* on a limestone hummock (urchin test 4 cm, depth 343 m, photo 119-91 Chave).

ern part of the island, the geological formations alternate between sand, talus, lava flows, hardpan, ash flows, and limestone reefs. Vast sand flats occur deeper than 1,800 meters off the Kona District and across the 'Alenuihāhā Channel between Hawai'i and Maui (Chave and Jones 1991; Young et al. 1992).

Lō'ihi Volcano

Lō'ihi is a submarine volcano located under the ocean about 16 kilometers southeast of the island of Hawai'i. Lō'ihi has grown from volcanic eruptions on a rift system that approximately follows the Hawaiian hot-spot trace on the ocean floor (Malahoff 1987). The highest point on Lō'ihi's summit is currently 950 meters below the surface. Several interdisciplinary teams currently study this emerging volcano using *Pisces V*. They have found pit craters and active vent areas on its summit that are similar to those found above sea level on other Hawaiian volcanos (fig. 24).

In the interior of the volcano are vertical dikes filled with hot rock or molten magma. Seawater, circulating freely through the porous pillow lavas and talus that cover the slopes of Lō'ihi, interacts with the hot dikes located there. The seawater becomes hot, leaches minerals from the surrounding basalt, and migrates upward as mineral-laden hot water (Malahoff et al. 1982; DeCarlo et al. 1983; Sedwick et al. 1992). The fluid flowing through the vents on the surface is 30°C when it comes into contact with the cold, 2°C seawater (fig. 25). The minerals that it holds precipitate out, mainly as iron oxide, a yellow to red powdery substance seen covering the summit (fig. 26).

Lō'ihi is built up of mechanically unstable talus with interspersed lava flows (Moore et al. 1982). Its summit and rift cones mark the sites of recent volcanic eruptions. Both its eastern and western slopes have slump scars where large masses of basalt have broken off and fallen downslope. Talus, composed of broken pillow and sheet lava, constantly moves down all the slopes. The submarine landslides on Lō'ihi generate earthquakes sufficiently powerful to be recorded on the seismometers at the Hawaiian Volcano Observatory (Klein 1982).

The southern end of the rift, the 90-million-year-old oceanic crust on which Lō'ihi sits, is fissured. Fresh rope-like (fig. 8) and pillow and tube lavas (fig. 11) have poured out of these fissures and covered the surrounding sediments (Malahoff 1992). North of the leading edge of the rift are 100-meter-high fresh pillow lava cones, only lightly dusted with sediment, that are sometimes colonized by animals (fig. 27).

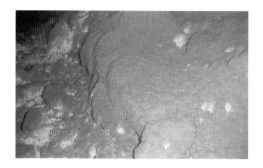

Figure 24. Numerous vents (white orifices) in Pele's Vents. Yellow hydrothermal iron oxide and bacterial mat covering ledge (photo area 4 m across, depth 970 m, photo 5085-95 Malahoff).

Figure 25. Shimmering caused by hot water emerging from vents and mixing with colder water (depth 990 m, photo 5027-62 McMurtry).

Figure 26. Inactive chimneys of hydrothermal iron oxide minerals (chimneys 10 cm, depth 1,015 m, photo 5222-12 Malahoff).

Pele's Vents, on the summit of Lōʻihi, is a major hydrothermal vent field that has been studied for almost a decade. Of the many instruments used to monitor this area, one of the most productive is the Ocean Bottom Observatory (OBO). Figure 28 shows a time-lapse video camera mounted on top of the OBO, a seismometer to the right, and temperature probes inserted in the individual vents. Fortunately, *Pisces V* had brought the OBO back to the surface before a huge section of Lōʻihi's summit (including Pele's Vents) collapsed in August 1996. Subsequent dives in this area of the volcano revealed a terrain vastly different from that previously explored.

Figure 27. Broken pillow talus slope with a glass feather sponge *Walteria leuckarti* between a red crinoid stalk and branched gorgonian *Chrysogorgia scintillans* (sponge 30 cm, depth 1,720 m, photo 5084-01 Malahoff).

Figure 28. Ocean Bottom Observatory at Pele's Vents (OBO length 0.8 m, depth 970 m, photo 5199-17 Malahoff).

Cross Seamount

Cross Seamount is in the central Pacific about 320 kilometers west of the island of Hawai'i. It became the site of extensive submersible studies between 1987 and 1993 because its geology is representative of the group of seamounts in the southern Hawaiian region that are older than the Hawaiian Islands. These seamounts were formed about 84 million years ago off South America on the site presently occupied by the East Pacific Rise and were carried by plate movement to their present location (Sager and Pringle 1987). The flanks of these seamounts are variously covered with ferromanganese crusts. Pinpointing the location of these crusts is of great practical importance because they contain cobalt, nickel, and platinum minerals.

Cross Seamount is star shaped and has a flat top that varies in depth from 350 to 500 meters. It drops at first steeply and then more gradually to the sea floor at a depth of 4,500 meters. The seamount's summit contains rounded boulders and worn basalt pinnacles that protrude through sandy areas, suggesting that the seamount may have risen above the surface in the past (fig. 29). Most of its original flanks have eroded through time, exposing massive elongated dikes that radiate out from the summit and down the flanks. Wide chutes filled with broken crust and talus extend from its summit to beyond its base. On its flanks are scarps and reworked, hardened debris called hardpan. Thick ferromanganese crusts cover the rock surfaces in many places (fig. 30). The slopes are probably stable over the life spans of most of the animals that attach to the rocks there. However, live animals are scarce in the chutes, where coral skeletons suggest that erosion and slumping of the seamount are active and ongoing (Malahoff et al. 1985).

Figure 29. *Walteria* sp., a glass finger sponge, attached to the side of a boulder protruding from a sand and pebble patch (sponge 30 cm, depth 1,260 m, photo 5036-122 Kelly).

Figure 30. *Narella bowersi* (left) and other primnoid gorgonians with brittle stars intertwined in their branches as well as a sea urchin on botryoidal manganese crust (sea urchin 10 cm, depth 970 m, photo 5141-27 Verlaan).

Moloka'i, Maui, and Lāna'i

The islands of Moloka'i, Maui, and Lāna'i are extinct volcanos formed by the Hawaiian hot spot a million years earlier than the island of Hawai'i (Clague and Dalrymple 1987).

Most submersible dives off Moloka'i explored the rim and slopes of Penguin Bank, the eroded summit of a sunken volcano to the west of Moloka'i. The top of this volcano is now a broad submarine shelf that extends to the island (Stearns 1966). This shelf is about 60 meters deep and is mostly covered by sand and coarse sediment. Around its perimeter from 60 to 200 meters' depth are limestone outcrops and sand patches, sometimes dotted with corals (fig. 31). At deeper depths, sandy slopes and eroded, limestone cliffs and benches are typical terrain. Sediment accumulates on the top of the bank, pours down the slopes, and covers the tops of most of the benches. Indeed, small sand avalanches were observed on almost every dive on the bank. Over time, the downslope movement of sediments wears deep grooves in the limestone. At a depth of 1,800 meters an extensive sediment plain abuts the bank. On this plain, lava from the sunken volcano is exposed as basalt outcroppings surrounded by sediment (fig. 32).

Recently, *Pisces V* dove off leeward Lāna'i, mapping fossil coral reefs and sediment-covered limestone terraces that were observed to depths of about 1,000 meters. *Makali'i* dove off Kahului, Maui, investigating a fine sediment plain at a depth of between 100 and 400 meters. The only irregular feature observed was a low limestone terrace that followed the 150-meter contour.

O'ahu

Submersible dives were concentrated off the southern shores of O'ahu in Māmala Bay off Honolulu and off the eastern point of the island. Many of the dives in Māmala Bay monitored sewer outfall effluents and surveyed dumped material dredged from Pearl Harbor. Whereas most of the bottom offshore from Pearl Harbor is debris laden (as seen behind the rakefish in fig. 33), most other areas of Māmala Bay are of fine sediment, relatively free of litter. The only large feature is a limestone bench that varies from 1 to 20 meters high and extends the length of the bay at a depth of between 350 and 400 meters.

Figure 31. Colonies of plate corals *Leptoseris* sp. on an old sponge-encrusted limestone reef (coral 25 cm, depth 107 m, photo 92-47 Ralston).

Figure 32. Stalked glass sponges *Caulophacus* sp. 3 and black coral *Stichopathes echinulosus (top)*, brisingid sea star *Freyella* sp. 2 *(middle)*, brachiopod *(bottom)*, and stalked crinoids *Ptilocrinus* sp. *(left)* and *Bathycrinus* sp. *(right)*, on a basalt boulder surrounded by fine sediment (sponge body 4 cm, depth 1,975 m, photo 5189-07 Mackenzie).

Figure 33. Cod *Laemonema rhodochir* in the background and rakefish *Satyrichthys engyceros* on a limestone bench. Dumped dredge spoil material is in the foreground (rakefish 12 cm, depth 367 m, photo 157-27 Sinton).

Beginning in 1982, scientists used *Makaliʻi* to study the vast Makapuʻu coral beds off Makapuʻu Point, southeastern Oʻahu. This area was discovered in the late 1960s, and precious pink corals were subsequently harvested. The coral beds are located on a 350-meter-deep terrace that drops abruptly to 500 meters and then descends into the channel between Oʻahu and Penguin Bank. This terrace and a shoreward series of shallower limestone terraces (fig. 15) were formed at sea level before Oʻahu sank to its present level (Stearns 1974). Basalt bombs ejected from Koko Head, a large tuff cone that erupted in prehistoric times (Bryan and Stephens 1993), are scattered across the terraces, some partially embedded in limestone.

The northern portion of the Makapuʻu coral beds has a limestone bottom swept clean of sediment (fig. 34), and the southern part of the beds is of limestone and coarse sediment. Limestone and sediment cover the western area of the beds (fig. 35), and to the east is an extensive plain of sand and fine sediment. Off Koko Head, where a few dives took place, tuff flows form submarine terraces that descend step-wise and terminate on a coarse sediment plain with limestone outcrops. Off Mōkapu Peninsula, to the east of Makapuʻu, the bottom is composed of basalt flows, eroded limestone, and sand.

Johnston Atoll

Johnston Atoll is a limestone-capped, sunken volcano unrelated to the Hawaiian Ridge. About 85 million years ago this volcano began to erupt and eventually emerged from the sea. In the ensuing millennia it became ringed with coral reefs, eroded, and sank. The atoll lies about 500 miles from its closest neighbor, French Frigate Shoals, an island in the Northwestern Hawaiian Islands. It is a typical low atoll composed of three barrier islands around a lagoon.

Makaliʻi made a series of dives around Johnston Atoll to depths of 400 meters (fig. 36). The ocean side of the atoll slopes steeply downward into an abyssal plain at a depth of 5,000 meters. The HURL bottom camera subsequently took photographs of basalt outcrops occurring far below the range of *Makaliʻi*. Keating (1987) discusses the geology of this atoll in detail.

From sea level to a depth of 40 meters, gentle sand and limestone slopes characterize the leeward side of the atoll, and reef corals occur in patches. On the windward side, storm waves and currents scour the bottom, and only a few colonies of small encrusting corals live in crevices. At depths from 40 to 400 meters, the atoll's limestone slopes

Figure 34. Primnoid *Callogorgia gilberti* and whip bamboo coral *Lepidisis olapa* on a limestone bench (primnoid 90 cm high, depth 413 m, photo 361-47 Grigg).

Figure 35. Branched bamboo gorgonian *Acanella* sp. 1 on limestone overlain by a thin cover of rippled sand (gorgonian 55 cm, depth 367 m, photo 149-01 Chave).

DEEP-SEA GEOLOGY 21

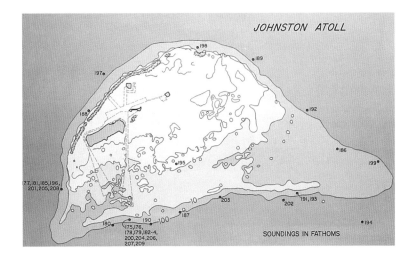

Figure 36. Map of Johnston Atoll showing submersible dive sites as red dots (depth contours in fathoms).

Figure 37. Group of deep-sea butterflyfishes *Chaetodon modestus* on a sand-draped, limestone ledge (fish 7 cm, depth 177 m, photo 188-03 Gooding).

are sediment draped, precipitous, and pitted by holes and caves (fig. 37). Most of the limestone benches are narrow, and broad terraces are rare. Fine sediment covers all flat features. The atoll's slope steepens between 120 and 400 meters, becoming almost vertical in some places. Its limestone walls, some smooth and some pitted, contain narrow ledges and large caves (fig. 38). Fine sand spills over the ledges, moves down chutes worn smooth by previous sand falls, and deposits on deeper benches (fig. 39).

Figure 38. Fluffy, ball-like neptheid corals on the roof of a limestone cave (neptheids 20 cm, depth 367 m, photo 199-55 Keating).

Figure 39. Two large solitary corals *Javania lamprotichum* on a limestone ledge (coral 5 cm, depth 330 m, photo 206-14 Eldredge).

2   *Deep-Sea Ecology*

COMMUNITIES OF MARINE ANIMALS can be found in Hawai'i from the highest tide marks of sandy or rocky shorelines to the greatest depths of the oceans. Excellent descriptions and photographs of Hawaiian marine life found at snorkeling or scuba depths are available in Randall (1985, 1996), Fielding and Robinson (1987), Hobson and Chave (1990), Hoover (1993), Severns and Fiene-Severns (1993), and Russo (1994). Our knowledge of animals that live at depths greater than 30 meters is not great, however, and few have been photographed. We can therefore cover only a limited number of Hawaiian deep-sea animal species in this book.

Many factors contribute to the distribution of deep-sea Hawaiian benthic, or bottom-dwelling, animals. These include species' origin, distribution, and degree of geographic isolation; light intensity; water chemistry, pressure, and temperature; current strength; and the nature of the bottom. The abundance of food, an extremely important condition, is related to all these factors (Kaufmann et al. 1989).

Origin, Distribution, Geographic Isolation

The young of most benthic marine animals drift or swim about in the plankton for a time before reaching a suitable place to settle. The vast stretches of ocean isolating the Hawaiian Ridge from other ridges and from Pacific Rim continents and island arcs are not linked by obvious current regimes. Yet, over time, larvae have been carried to Hawai'i from the eastern South Pacific and the western Indo-Pacific. Springer (1982), Newman (1986), and Wilson and Kaufmann (1987) discuss the different theories of the origins of Indo-Pacific marine fauna and how species came to Hawai'i. The consensus seems to be, however, that deep-sea larvae were probably carried along in currents coming from different directions and of different strength than those that transported shallow-water animals.

Of the Hawaiian species originating in the Indian and western Pacific Oceans, some had larval lifetimes long enough that they could be carried directly to the Hawaiian Islands in easterly currents. Those with shorter larval lifetimes first colonized islands or seamounts mainly located along the route from the northwestern Pacific to Hawai'i, their offspring reaching the Hawaiian Ridge some generations later. The ancestors of other marine species originated on the Pacific Plate when it was located off the coast of South America, and their larvae were carried between ancient, now sunken, islands and atolls as the plate moved toward the northwest, eventually reaching the Hawaiian hot spot. New species evolved as new Hawaiian hot-spot islands formed, and, as the Pacific Plate continued to move to the northwest and the islands began to sink into deeper waters, the animals living on the flanks began to disperse to the new islands formed by the hot spot.

Different kinds and different numbers of marine animals inhabit the Hawaiian chain than other Pacific areas. For example, the shallow-water crinoids (sea lilies) and gorgonians (sea fans) that are common elsewhere in the Pacific are not found in Hawai'i. Eldredge and Miller (1995) report about fifty-two hundred marine species inhabiting the shallow waters around the Hawaiian Islands, about one-third fewer than elsewhere in the Pacific. About 20 percent of these species are native to the islands (Chave and Kay 1985). In the deep sea around Hawai'i, about 35 percent of the species encountered by HURL submersibles are presently native to Hawai'i (calculated from table 1 at the back of this book). This calculation will probably change as more Hawaiian and Pacific deep-sea species are collected and identified.

From table 1 it can be seen that, of the sponge, coral, and echinoderm species now found in Hawai'i, 45 percent occur at depths of 15–400 meters, 15 percent at 400–800 meters, and 12 percent at 800–2,000 meters. Twenty-eight percent of the species in these groups range widely from 40 to 2,000 meters; those species that anchor to the bottom are often distributed in zones. Glass sponges, crinoids, and most gorgonians are found at depths greater than 300 meters; only a few gorgonians and black corals occur above 250 meters. Chave and Mundy (1994) report that the number of Hawaiian benthic fish species decreases logarithmically with depth. The greatest number of species inhabit depths between 15 and 200 meters, the fewest 2,000 meters. Crustaceans seem to follow the same distribution pattern as fishes.

Opposite:
Figure 41. Protected coral community showing two forms of *Porites* coral, yellow tang *Zebrasoma flavescens*, surgeonfish *Ctenochaetus strigosus*, damselfish *Chromis agilis*, and slate pencil urchins *Heterocentrotus mammillatus* (fish 9–10 cm, depth 15 m, photo Lobel).

Light Intensity

Many different kinds of plants and animals inhabit the Hawaiian reefs, and light intensity is one of the most important factors determining their distribution. Most important are the stony corals of which the reef is built up (fig. 7) and the coralline algae that cement coral debris, strengthening the reef (fig. 40)—both provide food and cover for other organisms in the reef community (fig. 41). Since minute symbiotic algae in the tissues of reef-building corals use sunlight to concentrate the nutrients in seawater that promote rapid coral growth, reef corals inhabit the shallower waters.

As depth increases, light intensity decreases, and the algae are eventually unable to grow. Reef-building corals and benthic algae therefore decline in number and size as depth increases from 30 to 100 meters

Figure 40. Coralline alga *Porolithon gardineri* growing on a coral reef (close-up, depth 10 m, photo Chave).

Figure 42. Deep reef with patches of coralline algae and a pair of *Chaetodon multicinctus* butterflyfishes (fish 6 cm, depth 30 m, photo Lobel).

(fig. 42). Below 100 meters, only a few, if any, small, stony corals dot the bottom (fig. 43). At these depths, non-reef-building corals and other animals obtain their food from the plankton, smaller animals, or dead animal and plant material floating in the water. Because deep Hawaiian waters are a nutrient desert, there is stiff competition for food. Filter-feeding organisms are able to capture more food particles in areas swept by faster currents, which are most often found on the sea floor moving over ridges, banks, and pinnacles (Genin et al. 1986).

Owing to low light intensity and nutrient-poor seawater, the depths below 500 meters are sparsely populated (Chiswell et al. 1990)—an

Figure 43. Hawaiian hogfish *Bodianus bilunulatus* near a limestone crevice (fish 45 cm, depth 122 m, photo 92-85 Ralston).

estimated 0.05 individuals per square meter (Chave and Jones 1991; Chave and Mundy 1994). Those animals that rely primarily on sight to obtain their food live in shallower waters, and, as twilight fades to darkness from 300 to 500 meters, many are found that have developed very large eyes. The species that live below 500 meters often have small eyes and well-developed, often bioluminescent body extensions—for example, fin rays or antennae—that are used to attract or sense food.

Water Chemistry

Salinity, temperature, and oxygen generally decrease, and pressure increases, with depth. Changes in these physical and chemical gradients may explain the distribution of many deep-sea animals. Water temperature, for example, varies between 23°C and 28°C at the surface, dropping abruptly to between 9°C and 12°C at the thermocline located between 100 and 300 meters (Chiswell et al. 1990), and reaching a constant 1°C–2°C at 2,000 meters. Many Hawaiian marine animals appear to have upper or lower depth limits that fall in the range between 100 and 300 meters, limits often associated with the thermocline boundaries. In Hawaiian waters, oxygen concentration is high (150 micromoles oxygen content) from the surface to about 300 meters. At 600 meters, oxygen concentration declines sharply to a minimum (20 micromoles oxygen content), and it is here that the most dramatic change in deep-sea fauna occurs.

Benthic Environment

The benthic environment also governs the abundance of deep-sea organisms. More species live on hard, irregular substrates than on smooth, hard surfaces. For example, the outcrop in figure 44, projecting above hardpan, contains tiny stalked crinoids, shrimps, and brittle stars, while the area around it is barren (HURL video data).

Many animals have adapted to a sandy environment by burrowing into it or growing stalks to keep their bodies above abrasive sand grains moved about by currents. They may also crowd onto isolated basalt outcrops standing above sand plains, settling mostly on the upper sections, where there is less abrasion and smothering by drifting sand (fig. 32). Some sessile forms attach to the upcurrent side of boulders. Others settle on the downcurrent side, shielded from currents

Figure 44. Two stalked crinoids *Ptilocrinus* sp., a shrimp, and a brittle star on basalt (shrimp 2 cm, depth 1,940 m, photo 5031-09 Jones).

Figure 45. Glass sponges *Farrea occa*, two gorgonians, black and white anemone *Actinernus* sp., and two flytrap anemones (hormathiid sp. 2) on a boulder embedded in sediment (hormathiid 8 cm, depth 1,652 m, photo 5016-04 Kerby).

Figure 46. Spiny crab *Lithodes longispinna* and flytrap anemones (hormathiid sp. 2) on a basalt boulder (anemones 6 cm, depth 1,260 m, photo 5059-03 Malahoff).

that can, potentially, sweep them away. The sponges, gorgonians, and anemones in figure 45 have settled on a boulder protruding from a sandy plain. Figure 46 shows a boulder on which large red spiny crabs perch and to which warty anemones are attached. These islands of life characterize the patchy distribution of animals on the deep ocean floor of the Hawaiian Archipelago.

The active areas of Lōʻihi volcano afford hostile environments for marine animals, which are discouraged from settling there by the noxious particles emitted from hydrothermal vents, sulfur-rich sediments, and unstable rocks. While large bacterial mats are present at the vent sites (Karl et al. 1988), only a few small animals live there, unlike the rich faunal assemblages inhabiting the older hydrothermal areas in the Pacific Ocean (Tunnicliffe 1992). Scientists have trapped tiny bresilid shrimp around the periphery of the vents (fig. 47), and occasional colonies of minute pogonophorans (*Sclerolinium* sp.) and other unusual animals can be found on the talus nearby (fig. 48). Occasionally fish swim by the vent areas, but none reside there.

Large animal species live on the slopes of Lōʻihi, away from the hydrothermal deposits and the vents. Older lava flows on the volcano's flanks, several hundred meters from the vent fields, sometimes host cnidarians (fig. 49) and large, sessile sponges. These communities are analogous to the communities found in the *kīpuka* of terrestrial Hawaiian volcanos—areas of older lava, usually covered by dense, mature vegetation and surrounded by newer, barren lava flows. Videotapes from the submersible *Mir* yielded sparse but persistent benthic life to a depth of 5,000 meters on the South Rift of Lōʻihi, but almost no animals were evident on the sediment-covered areas at the southernmost end of this rift zone.

Figure 47. A trap containing bresilid shrimp being collected by *Pisces V* (shrimp 3 cm, depth 980 m, photo 5242-78 Karl).

Figure 48. Siphonophore on a bacterial mat (body 5 cm, depth 980 m, photo 5106-44 Malahoff).

Figure 49. Gorgonians *Ellisella* sp. (right), *Metallogorgia melanotrichos* (center foreground, with a brittle star in its branches), and *Paragorgia dendroides* (center back) on a *kīpuka* of old pillow lava (*Ellisella* 35 cm, depth 1,910 m, photo 5145-05 Mahoney).

3   Deep-Sea Animals

Sponges (Phylum Porifera)

Most of the deep-sea sponges photographed by submersible cameras are different from the brightly hued, encrusting, and upright forms seen in shallow water. Hawaiian deep-sea varieties are usually white, yellow, or tan in color and have definitive body shapes composed of soft tissues and intertwined glass spicules (fig. 50). Table 1 lists the different types of sponges according to their taxonomic grouping. Those sponges described as red, pink, yellow, etc. are encrusting colonies found in fairly shallow water.

Some of the more spectacular deep-sea sponges have clear body tissues and intricate glass skeletons (fig. 51). Others look like crystal feathers (fig. 27) or transparent fingers (fig. 29). Sponges living on or near sandy patches have rounded or vase-shaped bodies raised above the bottom by long stalks of glass spicules (figs. 32, 52, and 53), an adaptation that lessens abrasion and clogging by sediment (Tabachnick 1991). Several deep-sea sponges adapt to different environmental conditions by varying their form as they grow. In figure 45 are four grayish white sponges identified as *Farrea occa*. On the downcurrent side of the boulder is a lettuce-like form, on the upcurrent side an upright branching form. Intermediate forms grow between the two anemones in the middle section of the boulder.

Sponges get their nourishment by sieving food particles from the water. Flagellated cells inside their body cavities beat continuously to draw in water through small pores, and amoeba-like cells lining channels inside the sponge catch the food particles that are pulled in. Water is expelled through larger pores, usually located on top of the sponge. Often the bigger openings are covered with sieve plates that prevent large particles from clogging the waterways (figs. 50 and 51).

Figure 50. Trio of glass sponges *Regadrella* sp. 1 on limestone (sponges 30–35 cm, depth 361 m, photo 150-06 Scheuer).

Figure 51. Glass bubble sponge *Corbitella* sp. on a basalt outcrop with primnoid gorgonians in the background (sponge 70 cm, depth 1,865 m, photo 5189-16 Mackenzie).

A variety of small animals live on or in sponges, probably because their hosts create currents that draw in food particles and because their spicules are sharp enough to deter most predators. Sometimes sponges are crowded with small crustaceans and brittle stars. The galatheid crabs on a vase sponge (fig. 54) and miniature shrimp and crab on a plate sponge (fig. 55) were observed catching and eating food particles drawn toward both sponges. Video footage of the plate sponge shows not only the shrimp and crab, but two fishes, a galatheid crab, a mollusk, and a brittle star as well. The plate sponge was collected, and twenty brittle stars, four shrimp, and about fifty worms were found in its tissues. Like their shallow-water counterparts, deep-sea sponges have their share of freeloaders. Tiny shrimp are often seen moving about the body cavities of glass sponges, and most sponges have at least one small crab or brittle star clinging to them (figs. 27, 29, 32, and 52).

Figure 52, *top left:* Stalked glass sponge *Bolosoma* sp. on cemented basalt talus (sponge 9 cm, depth 1,540 m, photo 5058-05 Jones).

Figure 53, *bottom left:* Stalked glass sponge *Trachicaulus* sp. (sponge body 35 cm, depth 1,820 m, photo 5146-01 Garcia).

Figure 54, *top right:* Long-armed galatheid crabs in a vase sponge attached to a basalt flow (sponge 18 cm, depth 850 m, photo 5074-95 Moore).

Figure 55, *bottom right:* Plate sponge *Corallistes*, crab *Dynomene devaneyi*, and two shrimp on a basalt boulder (sponge 30 cm, depth 346 m, photo 171-08 Chave).

Hawaiian glass rope sponges live in sandy areas where there is some movement of sediment downslope. Roughly half of these sponges have anemone-like zoanthids growing on their stalks (fig. 56). Many zoanthids are parasites, growing over and dissolving the tissues of their hosts. The zoanthids on rope sponges do not harm their hosts, perhaps because the sponge stalks are constructed of long glass spicule bundles. How zoanthids benefit from living on the sponge stalks or why they settle on only certain sponges remains a mystery.

Deep-sea sponges have numerous ways of anchoring to the substrate. The satin and porous glass sponges (fig. 57) and "scoop" glass sponges (fig. 58) attach to the bottom with tufts of tan spicules. On submarine ridges, where currents can be very strong, glass feather sponges (fig. 27) and finger sponges (fig. 29) attach to rocks with intertwined hook-like spicules. Colonies of enormous yellow stalked sponges (fig. 53) attach to current-swept ridges with long spicules resembling roots. The selected references on sponges listed in table 1 contain information about other attributes of these truly versatile animals.

Corals and Related Species (Phylum Cnidaria)

Deep-sea benthic cnidarians are sac-like animals called polyps. Individual polyps adhere to the sea floor and may live alone or develop into colonies of various sizes and shapes. Cnidarians have mouths ringed with tentacles that catch bits of food by secreting sticky or toxic substances and by firing specialized stinging cells. Some do not have skeletons, others have flexible internal skeletons, and others are encased in cups of limy material. Most are found where currents are strong and where there are many food particles in the water. Small forests of fan-, whip-, and bush-like cnidarians inhabit the current-swept ridges and pinnacles of central Pacific islands and seamounts (Grigg and Bayer 1976; Genin et al. 1986; Grigg et al. 1987; Chave and Jones 1991). Elsewhere, the creatures are sparse, occurring in patches on rocks or in the sand.

Figure 56. Glass rope sponges *Sericolophus* sp. in a fine sediment patch (sponge 14 cm, depth 422 m, photo 5302-97, France).

Figure 57. A porous sponge (Corbitellinae, undescribed genus and species) and a satin sponge (*lower right*), *Semperella* sp. 1, on a basalt slope (satin sponge 14 cm, depth 1,620 m, photo 5031-57 Zaiger).

Right: Figure 58. "Scoop" sponges, *Semperella* sp. 2, on a limestone bench (sponge 35 cm, depth 1,037 m, photo 5301-181 France).

Gorgonians are the most common group of cnidarians in the deep waters around the Hawaiian Islands (table 1). Very few live above the depth of 300 meters in the central Pacific, although they are common in the shallow waters of western Pacific islands and atolls. Members of the group are colonial and have horny internal skeletons and variously shaped spicules in their tissues. Most are flexible, stalked, and cemented to the bottom.

Primnoids are the most abundant gorgonian group in the Hawaiian Archipelago. Species are distinguished by the arrangement of their polyps, the color of their iridescent skeletons, colony shape, and skeletal spicules (Bayer 1981). Thus, the genus *Callogorgia* (figs. 34 and 59) can be distinguished from most other primnoid genera because it has pink polyps that face upward on the branches. When the covering tissue of *Calyptrophora agassizii* (fig. 60) is peeled away, its dark-gold skeleton distinguishes it from most other Hawaiian species in the genus, which have whitish gold skeletons. *Calyptrophora* species can usually be sorted by colony shape. A few having similar shapes are identified by the length of their stalks (figs. 61 and 62) or by other features. *Thouarella hilgendorfi*, a shaggy-looking primnoid with tiny

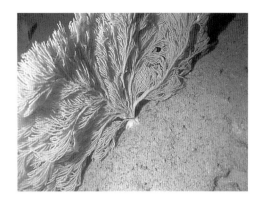

Figure 60. *Calyptrophora agassizii* on limestone (primnoid 110 cm, depth 367 m, photo 101-52 Scheuer).

Figure 61. Primnoids *Calyptrophora japonica* (*bottom*) and *Narella* sp. 1 (*upper left*) on limestone (*Narella* 30 cm, depth 367 m, photo 213-13 Chave).

Figure 62. Harp primnoid *Calyptrophora clarki* on limestone (primnoid 60 cm, depth 268 m, photo 96-69 Gooding).

Figure 59. A yellow gorgonian *Acanthogorgia striata*, *Callogorgia* sp. 1 (two branched), and *Narella* sp. 1 (fan shaped) on limestone (*Acanthogorgia* 120 cm high, depth 355 m, photo 54-24 Scheuer).

whitish purple polyps (fig. 63), is easily recognized, as is *Candidella*, whose large white polyps stick out at right angles from its branches (fig. 64).

Comb-like primnoids are easily distinguished from other primnoids by their widely spaced branches and small polyps (fig. 65). They are dominant off Moloka'i, have recently been collected, and are currently being identified. *Narella megalepis* has branches containing five to eight polyps per whorl (fig. 66), whereas many of the fan-shaped narellas have whorls of four polyps on their branches. Fan-shaped narellas dominate hard, smooth bottoms in several parts of the Hawaiian Islands. *Narella* sp. 1 (figs. 59 and 61) is the most abundant gorgonian in the Makapu'u coral beds (Grigg and Bayer 1976). *Narella* sp. 3 is the most common gorgonian on western slopes of the island of Hawai'i (Chave and Jones 1991). It resembles *Narella* sp. 1 but has more widely spaced polyps. *Narella bowersi* (fig. 30) is dominant on the flanks of Cross Seamount (Grigg et al. 1987).

Identifying other groups of gorgonians in their undersea habitats

Figure 63, *top left:* Thouarella hilgendorfi on limestone. Galatheid crabs are on its branches, and a hermit crab (*Parapagurus dofleini*) is near the top of the photo (primnoid 60 cm, depth 413 m, photo 148-30 Chave).

Figure 64, *bottom left:* Candidella helminthopora on basalt (*Candidella* 20 cm, depth 1,070 m, photo 5032-14 Jones).

Figure 65, *top right:* Comb-like primnoids on a basalt ledge (primnoids 60 cm high, depth 1,806 m, photo 5292-08 Holcomb).

Figure 66, *bottom right:* Narella megalepis on limestone (primnoid 50 cm, depth 398 m, photo 213-70 Chave).

can be difficult—and frustrating as well because dive time is limited. For example, having synthesized the complex biochemicals produced by blue-green paramuricids collected from the Makapuʻu coral beds, scientists found that one specimen produced a chemical that successfully treats a certain form of cancer (Okuda et al. 1982). However, when additional specimens were needed to complete biochemical screening, in order to isolate the correct species all blue-green paramuricids had to be collected and reidentified since blue, blue-green, and green paramuricid species (fig. 67) cannot be distinguished one from another until their spicules are examined under a microscope. As luck would have it, most of the easily identified gorgonians (e.g., *Bebryce* in fig. 68 and *Acanthogorgia* in fig. 59) do not produce compounds that are useful for pharmaceutical purposes.

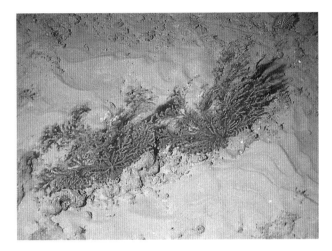

Left, top to bottom:
Figure 67. Two blue paramuricid gorgonians on limestone covered with a thin layer of rippled sand (gorgonians 30 cm, depth 339 m, photo 66-08 Scheuer).

Figure 68. *Bebryce brunnea* on limestone with scattered remains of dredged coral skeletons (*Bebryce* 50 cm, depth 355 m, photo 66-46 Scheuer).

Below:
Figure 69. Chrysogorgiid gorgonians *Chrysogorgia stellata* (bushy) and *Iridogorgia superba* (coiled) on a pillow lava flow (*Iridogorgia* 20 cm, depth 1,530 m, photo 5147-11 Mahoney).

Figure 70. Feathery chrysogorgiid on basalt talus (chrysogorgiid 15 cm, depth 1,870 m, photo 5145-73 Mahoney).

Figure 71. *Iridogorgia bella* on a basalt slope (gorgonian 25 cm, depth 1,775 m, photo 5087-23 Malahoff).

Figure 72. *Acanella* sp. 2 on a breadcrust pillow (gorgonian 20 cm high, depth 1,910 m, photo 5145-32 Mahoney).

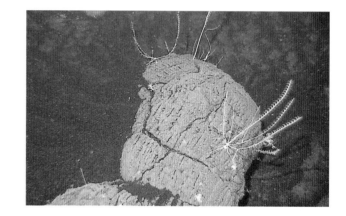

Figure 73. *Keratoisis* sp. 4 on a basalt ledge (bamboo gorgonian 150 cm, depth 1,725 m, photo 5087-65, Malahoff).

Chrysogorgiids are among the most beautiful and unusual of gorgonian groups. The lower parts of their stalks are usually bare and iridescent. Branching, coiled, fluffy, and feathery species occur in patches throughout the Hawaiian Islands from depths of 250 to over 2,000 meters (figs. 27, 69, and 70). An umbrella-like species is the most abundant of Hawaiian chrysogorgiids and has the greatest depth range (table 1). This animal has well-spaced pink polyps on branches that radiate horizontally from the top of its stalk (fig. 49). The strangest chrysogorgiid so far encountered resembles a dandelion seed cluster (fig. 71). The entire colony of polyps grows on branches that cascade from a tightly coiled stem.

The skeletons of bamboo corals are composed of alternating brown horny nodes and white calcareous segments. The nodes give these gorgonians some flexibility. Members of this group have numerous growth forms. The whip bamboo coral grows as a single, coiled stalk that may spiral to a height of three meters above the ocean floor (fig. 34). The orange fan-like species in figure 35 branches from the horny nodes, as does the stick-like form in figure 72. A common bush-like species on Cross Seamount branches from its white segments (fig. 73). Most bamboo corals are bioluminescent. When touched, they emit pulses of greenish light that travel away from the contact point along the length of the colony. If time permitted when bamboo corals were encountered, the submersible pilots liked to turn off the lights and touch these animals so that the observers could have a firsthand look at gorgonian pyrotechnics.

The skeletons of pink corals are packed with hard, fine-grained spicules suitable for cutting, carving, and polishing into jewelry. Precious corals range in color from white to pink (fig. 74). Usually, specimens collected in deeper waters are darker than those collected in shallower waters. Two precious corals, dubbed by jewelers *shocking pink* (fig. 75) and *angelskin* (fig. 76), were dredged from the Makapu'u beds in the early 1970s (Grigg 1977). Their numbers were quickly depleted, however, and harvesting operations ceased. Fifteen years later, submersible studies at Makapu'u showed that precious coral growth and recruitment were proceeding very slowly (Grigg 1988). The harvesting of precious corals in Hawai'i is now limited to selective collection by submersibles.

The distribution of Hawaiian deep-sea gorgonians is not well understood because very little of the sea floor has been surveyed. For example, two species of *Paragorgia*, a white fan form (fig. 77) and a fleshy form (fig. 78), attach to both limestone and basalt, but a red

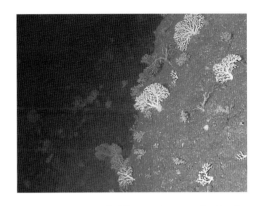

Figure 74. Group of white precious corals *Corallium secundum* and yellow dendrophyllid corals on a vertical limestone cliff (dendrophyllid 10 cm, depth 190 m, photo 96-57 Gooding).

Figure 75. Shocking pink precious coral *Corallium regale* on limestone (coral 20 cm, depth 413 m, photo 148-85 Chave).

Figure 76. Angelskin precious coral *Corallium secundum* on limestone (coral 30 cm, depth 367 m, photo 146-16 Scheuer).

species (fig. 49) is found only on basalt. What is not known is whether the larvae of the red paragorgiid choose only basalt as a substrate or whether they attach to limestone as well and have simply never been observed in such an environment.

In the deep waters of the Hawaiian Islands and Johnston Atoll, cnidarian species belonging to the groups Hydroida, Milleporina, and Stylasterina are usually found in small colonies cemented to rocks. Only a few, like the violet coral, are large enough to be photographed from submersibles (fig. 79). Colonies of violet corals live on the seaward side of Johnston Atoll's reef platform and dot the outer slope to depths of about 200 meters. They are common elsewhere in the Pacific but do not occur in the Hawaiian Archipelago.

Large bottom-dwelling comb jellies (Phylum Ctenophora) are also present in the Hawaiian Archipelago and Johnston Atoll. These creatures attach to hard substrates or dead cnidarian stalks at depths of between 300 and 400 meters (fig. 80). Comb jellies have sac-like bodies and two extremely long, retractable, comb-like tentacles used for feeding. Unlike true jellyfish, they capture their prey with sticky secretions rather than stinging cells. Animals that were knocked off dead whip corals on one of *Makali'i*'s dives kept reappearing on the same stalks on subsequent dives. We were never able to discover how these animals reoccupy the stalks.

Soft corals (Alcyonacea) frequent the Hawaiian Islands and Johnston Atoll. The roof of a huge limestone cave 400 meters below the surface of Johnston Atoll was almost completely covered with ball-like

Figure 77. *Paragorgia regalis* on limestone (gorgonian 80 cm, depth 355 m, photo 282-19 Scheuer).

Figure 78. *Paragorgia* sp. on limestone (gorgonian 40 cm, depth 398 m, photo 213-72, Chave).

Figure 79. Limestone cave fringed with colonies of violet coral *Distichopora violacea* (*right*) and orange sponges. It is inhabited by soldierfishes *Myripristis chryseres* and squirrelfishes *Neoniphon aurolineatus* (fishes 20 cm, depth 145 m, photo 193-49 Ralston).

neptheid colonies (fig. 38). Why these animals crowd together this way remains a mystery, although it is suspected that they do so in order to take advantage of the intensified water flow. Other soft coral colonies attach to rocks and are often the first to colonize artificial substrates (Moffitt et al. 1989). The toadstool coral attaches to large talus blocks and other types of hard substrates (fig. 1). It varies in color from orange to red and has a few widely spaced polyps that can be pulled into the upper portion of its body. The area surrounding the retracted polyps is white, and the overall effect is that the coral resembles the typical red toadstool with white spots commonly seen in illustrations. A tan coral, similar to the toadstool coral, is usually surrounded by small anemones with very large bases (fig. 81). The relationship between the two species is unknown.

In some areas, colonies of polyps connected by tube-like stolons (Stolonifera, fig. 82) cling to the bottom with root-like threads. Most of these animals were too small to be photographed.

Most Hawaiian sea pens (Pennatulacea) live in sediment, one exception being the brown rock pen (fig. 82). The base of the rock pen is unusual, functioning like a suction cup to allow the animal to attach itself to hard substrates. The rock pen is bioluminescent, as are most of its relatives, but it often must be touched several times before it emits golden flashes of light. Most sea pens wedge themselves in sand or fine sediment by means of expandable bulbous stalks (fig. 83). Some, like the white virgularid sea pen (fig. 84), are able to pull their bodies completely into the sediment. The submersible's lights and the water motion caused by the thrusters did not affect these sea pens, but, when the manipulator attempted to collect them, the animals retracted.

Single polyps such as the Corallimorpharia (fig. 85) and sea anemones (Actiniaria) attach to rocks with suction cup–like bases. Unfortunately, soft-bodied anemones are difficult to collect in deep water, and little is therefore known about them. Long-tentacled anemones (fig. 86) inhabit cliff faces. Close-up video images of one of these animals showed tiny shrimp-like animals stuck to its tentacles. Presumably, the anemone pulls the tentacles into its mouth and removes the food particles.

Figure 80. Creeping comb jellyfish (*Lyrocteis* sp.) attached to a basalt pillow (comb jelly 4 cm, depth 361 m, photo 112-21 McMurtry).

Figure 81. Alcyonacean *Anthomastus steenstrupi* surrounded by wide-based anemones on hard pan (alcyonacean 9 cm, depth 975 m, photo 5082-07, Young).

Figure 82. Rock pen *Calibelemnon symmetricum*, angelskin precious coral (*Corallium secundum*), and purple stoloniferans *Clavularia* sp. 1 on limestone (rock pen 4 cm, depth 413 m, photo 148-57 Chave).

Figure 83. Sea pen *Funiculina* sp. buried in sand (sea pen 16 cm, depth 367 m, photo 215-16 Chave).

Figure 85. *Corallimorphus* sp. on manganese-encrusted basalt (cnidarian 50 cm, depth 817 m, photo 5093-88 Malahoff).

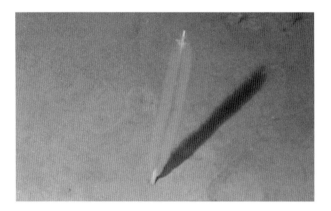

Figure 84. Sea pen *Virgularia* sp. and two cerianthids in fine sediment (sea pen 14 cm, depth 251 m, photo 99-04 Chave).

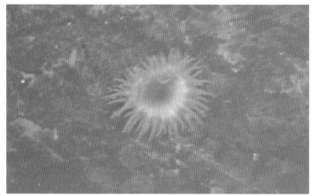

Figure 86. A long-tentacled anemone on a basalt ledge (tentacles 12 cm long, depth 915 m, photo 5015-33 Kerby).

Figure 87. A flytrap anemone (hormathiid sp. 3), cnidarians, tubeworms, and brittle stars on a basalt outcrop. A hermit crab in a scaphopod shell is on the sand near the outcrop (anemone 10 cm, depth 1,800 m, photo 5080-06 Young).

Figure 88. Three flytrap anemones *Actinoscyphia* sp. 3 clinging to a primnoid gorgonian and a black and white anemone on a basalt boulder (anemone 4 cm, depth 1,260 m, photo 5062-60 Malahoff).

Several deep-sea anemones resemble Venus's-flytrap plants. Some attach to living or dead animal stalks, others to rocks and debris such as paint cans and wooden blocks. Figure 87 shows one of these anemones with very large knobby warts clinging to a basalt boulder. A beer can lying near the rock indicates the dimensions of the animal. Figure 88 shows three flytrap anemones clinging to a gorgonian and a black and white anemone attached to a rock. For unknown reasons, black and white anemones are almost always found with other types of large anemones (figs. 45, 88, and 89).

By far the strangest member of the group is the pink flower anemone (fig. 90). This animal is soft bodied and possesses simple, easily detached tentacles. When the submersible attempted to collect the animal, tentacles flew in all directions, and there was practically nothing left to capture.

The black corals (Antipatharia), another common deep-sea group, have fine-grained ebony skeletons suitable for jewelry manufacture. Underwater, the living tissue around the skeletons of black corals may be variously colored. The shiny black coral with its red polyps and glossy black skeleton is one of the most beautiful of the group (fig. 91).

Black corals are usually observed attached to ridges, benches, and hummocks where currents are strong. The black coral with blue-gray polyps in figure 92 and the smaller black corals with white polyps on a few flexible, curving branches in figure 23 live on limestone outcrops off the east shore of the island of Hawai'i, south of Hilo. Many lava tubes on the slopes of this island carry freshwater from the land into

Figure 89. Black and white anemone *Actinernus* sp. and actinostolid anemone on two boulders (actinostolid 12 cm, depth 1,490 m, photo 5080-22 Young).

Figure 90. Flower anemone *Liponema brevicornis* and a small ruffled coral *Balanophyllia laysanensis* on a sand-dusted talus slope (anemone 12 cm, depth 336 m, photo 59-49 Wiltshire).

Figure 92. Black coral *Antipathes intermedia* with brown and blue zoanthids overgrowing it. A cidarid urchin *Histocidaris variabilis* is in its branches, and a grenadier (fish) is behind it (urchin test 6 cm, depth 379 m, photo 220-05 Gooding).

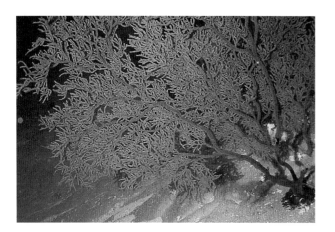

Figure 91. *Leiopathes glaberrima* on sediment-dusted limestone (black coral 140 cm, depth 297 m, photo 199-03 Keating).

Figure 93. *Antipathes ulex* on the edge of a freshwater aquifer flowing through a lava tube (black coral 180 cm, depth 208 m, photo 224-77 McMurtry).

Figure 94. Dark and light varieties of black wire corals *Cirrhipathes spiralis* on limestone covered with a dusting of sand (corals 80–140 cm, depth 171 m, photo 96-30 Gooding).

Figure 95. Feathery black coral *Bathypathes conferta* on limestone (part of coral showing 90 cm, depth 382 m, photo 103-23 Chave).

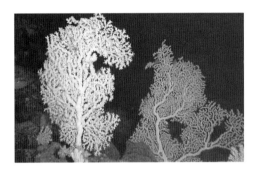

Figure 96. Gold coral *Gerardia* sp. on pillow lava (corals 100 cm, depth 355 m, photo 5078-54 Moore).

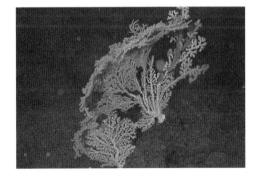

Figure 97. Gold coral *Gerardia* sp. overgrowing *Acanella* sp. 1, a branched bamboo gorgonian attached to pillow lava (largest coral 60 cm, depth 346 m, photo 114-140 Epp).

the ocean. An attractive species of feathery black coral grows near the mouths of these tubes (fig. 93) and also occurs along basalt ledges.

Wire corals frequent limestone slopes. There are two forms of wire coral, one with small brown polyps and one with large white polyps (fig. 94). Although they look very different underwater, the two forms apparently belong to the same species. One of the largest of the black corals resembles a feather and can reach four meters in length (fig. 95). This species varies in color from dark red to whitish tan, and its stalk is extremely tough and flexible. Its base is so securely anchored to the bottom that specimens could be collected only after the protein material cementing the stalk to the limestone was scraped away.

Deep-sea zoanthids are colonies of soft-bodied polyps that grow over other cnidarians instead of making their own supporting skeletons. They are parasitic, feeding on the polyps of their cnidarian hosts. The stinging cells and toxic substances of most deep-sea cnidarians do not harm human skin, but zoanthids are handled with gloves, preventing painful rashes.

At depths of 360 meters and below, gold corals (parazoanthids with iridescent gold skeletons and large yellow polyps) attach to hard substrates (fig. 96). The skeletons of these animals are sometimes used to make jewelry (Grigg 1977). There may be a large number of gold coral species, but many are very similar to one another, and most appear identical underwater. In Hawai'i, the larvae of gold corals locate and settle on bamboo corals (fig. 97).

Zoanthids may produce pharmaceutically important compounds as well as compounds that are bioluminescent and fluorescent. Figure 98 shows a bioluminescent species of zoanthid that has grown over and digested the tissue from all but two branches of the pink finger primnoid. In figure 92, the large polyps of the blue zoanthid are overgrowing part of a black coral, and a purple cidarid urchin is eating the zoanthid polyps. Brown hydroids and tiny white barnacles attach to the dead parts of the black coral branches.

Members of the stony corals (Scleractinia) secrete dense skeletons of calcium carbonate and often form large colonies in shallow water, some building vast reefs to depths of 30 meters (fig. 7). From 30 to 170 meters only a few species of plate-like reef-building corals live on Hawaiian island slopes (fig. 31). At about 120–180 meters, small white stony corals may become populous on slopes (fig. 99). These corals feed on plankton in the water column.

Below 180 meters there are places where small colonies of yellow dendrophyllid corals inhabit limestone cliffs (fig. 74). These corals are part of an assemblage that includes a harp primnoid (fig. 62), a small banded crinoid, and white and pink angelskin corals. These assemblages occur off Penguin Bank, eastern Maui, French Frigate Shoals, and perhaps New Caledonia (see Roux 1994) and differ only in species abundance: off Penguin Bank there are fewer crinoids and spiral black corals and more precious corals, harp primnoids, and yellow dendrophyllid corals than off Maui and French Frigate Shoals.

Deeper than 250 meters, stony corals are small and dot the landscape, especially on older lava flows and limestone formations. Most are solitary, but some species may form small colonies. The small tube corals and the ruffled coral in figure 90 are single animals that usually attach to basalt. Figure 100 shows several solitary stony corals belonging to a species that lives on sediment. The animal in the right-hand corner of the photograph has expanded its delicate white tentacles. One of the most beautiful of the deep-sea corals has a flaring white skeleton and dark red polyps (fig. 39). This animal frequents smooth limestone benches and may be found to depths of 400 meters.

Figure 98. A primnoid gorgonian *Narella megalepis* on limestone. *Parazoanthus* is growing over the gorgonian (primnoid 35 cm, depth 367 m, photo 213-08 Chave).

Figure 99. Branched stony corals *Madracis kauaiensis* on an eroded limestone reef (corals 8 cm, depth 183 m, photo 96-87 Gooding).

Figure 100. Solitary corals *Dendrophyllia serpentina* and a crab *Cyrtomaia smithi* on fine sediment (crab carapace 4 cm, depth 349 m, photo 119-78 Chave).

Sea Stars and Related Species (Phylum Echinodermata)

Echinoderms are another large group of invertebrates inhabiting the deep water around the Hawaiian Islands (table 1). Two striking features set echinoderms apart from other phyla: a water vascular system used for locomotion and body parts that are arranged outward like spokes on a wheel. Some deep-sea echinoderms resemble their shallow-water counterparts, but others are very unusual.

Sea stars (Asteroidea) have flexible bodies with five or more arms. Sand dwellers, like *Calliderma spectabilis* (fig. 101), dot fine sediment plains located near the bottom of steep precipices. These sea stars are part of a community including anemones, long-spined sea urchins, and sea pens that exists at depths of between 150 and 300 meters on sediment where currents are minimal. The leaf star (fig. 102) moves over the sand at a very slow pace. When the animal was first seen, observers thought it was a leaf—until it was turned over. In response to being disturbed, it immediately rolled itself into a tube, quickly righting itself and flattening out on the sand again. At greater depths, multirayed sea stars (fig. 103) occur on large tracts of coarse sand. They glide rapidly away from the submersible's light path on hundreds of tube feet and bury themselves under the sand if pursued.

Most sea stars encountered moved more slowly on hard substrates. During the start of a dive in the Makapuʻu gorgonian beds, a thick-bodied granular sea star was observed gliding away from a large boulder. After 20 minutes it had traveled about 20 centimeters, reaching the small brown rock pen seen in figure 104, and had moved only

Figure 101. Sea star *Calliderma spectabilis* and long-spined urchin *Chaetodiadema pallidum* on fine sediment (urchin test 6 cm, depth 213 m, photo 211-36 Chave).

Figure 102. Leaf star *Anseropoda insignis* on fine sediment (leaf star 16 cm, depth 260 m, photo 124-08 Maragos).

Figure 103. *Solaster* sp. on coarse dark sand (sea star 26 cm, depth 975 m, photo 5082-11 Young).

Figure 104. *Cryptopeltaster* sp., a brown rock pen, and small gorgonians on limestone (sea star 25 cm, depth 382 m, photo 148-07 Chave).

2 meters farther on when the submersible returned 4 hours later. Between 250 and 1,000 meters' depth, ornate sea stars often occur on limestone near or in limestone crevices (fig. 18). Deeper-dwelling sea stars cling to basalt structures with their tube feet. The two shown in figures 105 and 106 moved away very slowly when prodded by the submersible's manipulator.

Sea stars feed on a variety of invertebrates. *Coronaster,* an orange multiarmed sea star, lives on sediment populated by large colonies of tiny brittle stars. Dotted about the brittle star beds are barren spots called halos, in the centers of which are sometimes found multiarmed sea stars (fig. 107). Several of these sea stars were collected, and all had stomachs full of brittle stars. Either *Coronaster* had eaten all the brittle stars in the halos, or the brittle stars had moved away from these predators, or both. *Calliaster* (fig. 108), seen near the base of a gorgonian stalk at the beginning of one dive, was found a half meter up the stalk 3 hours later. A half meter of gorgonian polyps had been eaten, and their distinctive skeletal spicules were later found in the sea star's stomach. *Pentaceraster cumingi* often occurs in pen shell beds (fig. 109), and, when they are collected, their stomachs are often everted and wrapped around half-digested mollusk shells.

Large brisingid sea stars perch on ledges of steep slopes where currents are strong (fig. 110). Brisingids are sea stars with arms and tube feet modified so that they can actively capture their food from the water column. Emson and Young (1994) report that they saw one of the brisingid species catching and eating small swimming crustaceans.

The deep-sea urchins (Echinoidea) are similar to their shallow-water relatives. Their bodies are enclosed in a rounded test through which tube feet and spines extend. Sea urchins inhabit sandy areas or smooth hard substrates. Some, especially the irregular urchins, move so rapidly that they are almost impossible to collect, and most glide rapidly away from the glare of submersible lights.

Tan-colored long-spined urchins are abundant on the sediment plains (fig. 101). The urchins moved rapidly away when they were touched or when the submersible lights were turned on. Despite the large number of animals in the area, very few urchin tracks were observed. Apparently they do not move about very much under normal conditions and perhaps feed on material caught between their spines.

A red echinothurid urchin inhabits the sandy areas populated by tiny brittle stars and *Coronaster.* This urchin is also found in the

Figure 105. *Asterodiscides tuberculosus* on a basalt dike (sea star 6 cm, depth 1,375 m, photo 5182-01 Garcia).

Figure 106. *Hymenaster pentagonalis* on basalt (sea star 18 cm, depth 1,800 m, photo 5253-17 Moore).

Figure 107. *Coronaster eclipes* and an aggregation of brittle stars (*Histampica cythera*) on fine sediment (sea star disk 4 cm, depth 252 m, photo 124-54 Maragos).

Figure 109. A sea star *Pentaceraster cumingi* on a bed of pen shells (*Pinna muricata* shells 12 cm, depth 52 m, photo 110-03 Ralston).

Figure 108. *Calliaster pedicellaris* on the gorgonian *Ellisella* sp. (sea star 20 cm, depth 1,740 m, photo 5080-13 Young).

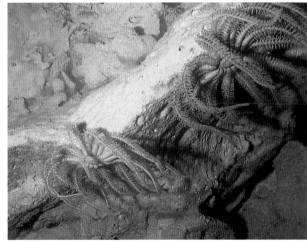

Figure 110. Brisingid sea stars clinging to an outcrop (sea stars 23 cm, depth 360 m, photo 43-16 Colin).

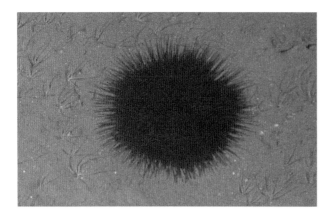

Figure 111. An echinothurid urchin in the middle of a brittle star aggregation in fine sediment (urchin 12 cm, depth 232 m, photo 125-40 Devaney).

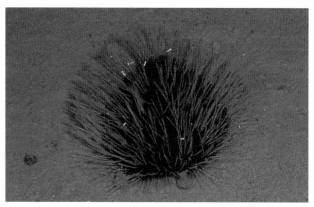

Figure 113. *Phrissocystis multispina* on fine sediment (urchin 7 cm, depth 1,800 m, photo 5080-08 Young).

Figure 112. *Phormosoma bursarium* on fine sand (urchin 9 cm, depth 530 m, video 5225 Young).

Figure 114. Cidarid urchin *Actinocidaris thomasii* on black coral *Antipathes subpinnata* (urchin test 5 cm, depth 214 m, photo 199-84 Keating).

Figure 115. Aggregation of cidarid urchins *Prionocidaris hawaiiensis* on coarse carbonate sediment (urchin test 4 cm, depth 92 m, photo 108-07 Ralston).

middle of halos (fig. 111). It probably eats brittle stars, but it is an extremely active animal and has so far defied capture, much to the dismay of pilots and scientists. At greater depths, bladder urchins scuttle across the sand on specialized flattened spines. On top of this urchin's test are spines modified into bladders of unknown function (fig. 112). Still deeper, a large reddish brown urchin (fig. 113) frequents sandy areas. This urchin also moves very rapidly on top of or buries itself in the sand if disturbed. It has hollow spines and a very fragile test, both features that perhaps allow it to walk on soft sediment without sinking in.

Several species of cidarid urchins on Hawaiian undersea slopes live as far down as depths of about 400 meters. The most common of these is a brown cidarid that usually clings to hard substrates and occasionally climbs on black coral (fig. 114). A purple cidarid (fig. 92) is also found on black corals. This species eats zoanthids, but the diet of the brown-spined cidarid is unknown. *Prionocidaris hawaiiensis* was commonly encountered only on the periphery of Penguin Bank. The large number of animals in figure 115 perhaps represents a feeding front or spawning aggregation (see Billett and Hansen 1982). The group of white cidarids clustering on a limestone outcrop in figure 23 is perhaps a spawning aggregation (see Young et al. 1992). Echinoderms often aggregate so that their sperm and eggs are in close proximity when shed into the water.

Clusters of skunk urchins cling to smooth limestone benches where there are steep cliffs above and below. If currents are strong, they turn so that their spines point into the current (fig. 116). Perhaps this orientation enables them to trap more food particles between their spines. The submersible pilots dubbed this species skunk urchin because of its black and white coloration.

Long-spined urchins are grazers. Their long spines are used for protection and to detect water movement; smaller spines underneath their bodies are used to move along the ocean floor. Often groups of colorful long-spined urchins aggregate on steep slopes, especially at Johnston Atoll (fig. 117). Occasionally closely spaced individuals were observed shedding milky sperm and eggs into the water. Another type of long-spined urchin (fig. 118) is solitary and lives in much deeper water. It is a very small animal with extremely long spines on the top of its body. This species moves very rapidly across the bottom on its shorter spines.

Most sea cucumbers (Holothuroidea) are to be found on the sea floor, slowly plowing the sand as they search for food particles. Most swim by undulating their bodies, and some spend most of their lives swimming and drifting in the water column (Billett and Hansen 1982). Some deep-sea species (fig. 119) have cigar-shaped bodies similar to those of shallow-water forms, while others display unusual body shapes and appendages (fig. 120). The different species of sea cucumbers living at deeper depths have a variety of defense mechanisms. A speckled sea cucumber resembles the coarse sediment on which it lives (fig. 121) and rapidly burrows into the sand when touched. A species of large purple sea cucumber that stands out in its environment (fig. 122) produces a noxious compound in its skin that may make it inedible. As are many deep-sea cucumbers, this species is gelatinous and very light. When touched, individuals usually push themselves off the bottom and undulate upward into the water column.

Deep-sea brittle stars (Ophiuroidea) were usually seen clinging to gorgonians (fig. 123) and sponges. Sometimes, when the submersible disturbed the sediment, tiny brittle stars floated out into the water. Judging from the size of the brittle star population where *Coronaster* and red echinothurids occur, brittle stars may be very common in soft

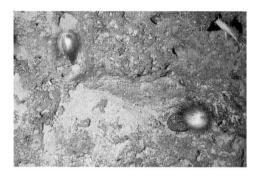

Figure 116. *Eurypatagus ovalis* on an eroded limestone cliff (urchin 14 cm, depth 176 m, photo 289-31 Moore).

Figure 117. Group of long-spined sea urchins *Diadema savignyi* and stylasterine corals on a steep limestone slope (urchin test 6 cm, depth 171 m, photo 187-40 Ludwig).

Figure 118. *Aspidodiadema arcitum* on a sand-covered ledge of a basalt outcrop (sea urchin test 3 cm, depth 1,985 m, video 5192 Bertram).

Figure 119. *Bathyplotes patagiatus* on iron oxide–rich sediment (sea cucumber 12 cm, depth 1,560 m, video 5198 Cremer).

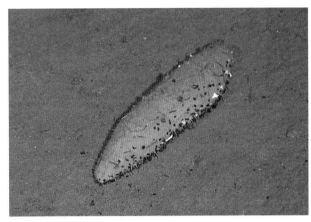

Figure 121. *Mesothuria carnosa* on fine sediment (sea cucumber 25 cm, depth 1,775 m, photo 5080-12 Young).

Figure 120. *Benthodytes* sp. on sand (sea cucumber 30 cm, depth 1,900 m, video 5189 Mackenzie).

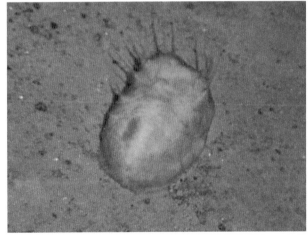

Figure 122. *Paelopatides retifer* on fine sediment (sea cucumber 15 cm, depth 1,060 m, photo 5082-03 Young).

sediments (fig. 111). In shallow water, brittle stars usually remain under rocks, but deep-sea forms are often observed in the open. Numerous orange brittle stars were seen on talus blocks off windward Hawai'i (fig. 124). They moved between and under the rocks when the submersible's bright lights were turned on.

Basket stars, close relatives of brittle stars, are known as gorgonocephalids because their coiling branched arms move sinuously and resemble the snaky hair of Medusa, one of the three Gorgons of Greek mythology. Groups of basket stars were seen clinging to old pillow lavas off the island of Hawai'i and on the flanks of Lō'ihi. They were also seen on dead gorgonian stalks off O'ahu and Cross Seamount (fig. 125).

There are no shallow-water sea lilies (Crinoidea) in the Hawaiian Archipelago, but deep-sea species abound. Hawaiian crinoids range from stalked animals permanently anchored to the hard bottom to those that can crawl or swim using their feathery arms. In between these two extremes are stalked crinoids that creep about using the cirri on their stalks and unstalked species that cling to skeletons of sessile animals and hardly move at all.

Unstalked crinoids often perch on dead sponges (figs. 126 and 127) or cnidarians (figs. 128 and 129). Grasping with hook-like cirri, they spread out their feather-like arms to trap food particles from the water. Brown comatulid crinoids are extremely motile. They can be found on hard substrates, most commonly talus, holding on to irregularities in the rock with their cirri, their arms extended upward. If touched, these animals swim gracefully upward and then glide downward to land in another spot. On first landing after being disturbed, they flatten out completely (fig. 130) and then slowly raise their arms again. Many beautiful crinoids remain unidentified because they have not yet been collected. The lucky yellow animals in figure 131 are probably still hanging from their ledge because the submersible's collection basket was full when they were found.

The most spectacular crinoid encountered by *Pisces V* is a red stalked animal that commonly occurs on the slopes of Lō'ihi volcano, Cross Seamount, and the island of Hawai'i. This large species has very broad feathery arms that fan out in all directions (fig. 132). Many of the Hawaiian stalked crinoids are tiny; often they either go unnoticed or are mistaken for other animals. The small, yellow, stalked crinoids in figures 32 and 44 were often noticed only during postdive data analyses of videotapes. The tiny white crinoid in figure 32 was often mistaken for a tubeworm.

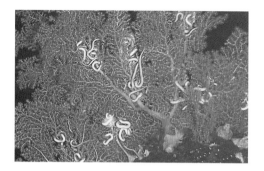

Figure 123. Two types of brittle stars intertwine in the branches of blue paramuricid gorgonians (gorgonian 36 cm, depth 785 m, photo 5093-113 Kelly).

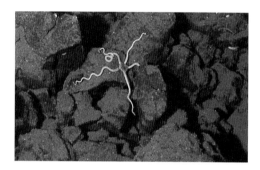

Figure 124. *Ophiomyxa fisheri* on a basalt talus flow (brittle star disk 1 cm, depth 349 m, photo 112-17 McMurtry).

DEEP-SEA ANIMALS 51

Figure 125. Dead *Pleurogorgia* stalk with two gorgonocephalid basket stars (basket star disks 5 cm, depth 1,335 m, photo 5091-69 Kelly).

Figure 127. Charitometrid crinoid on a dead sponge attached to a *pāhoehoe* flow (crinoid 16 cm, depth 920 m, photo 5059-38 Zaiger).

Figure 128. Crinoids *Cosmiometra crassicirra* perched on a black coral stalk on limestone (crinoids 5 cm, depth 257 m, photo 125-30 Devaney).

Figure 126. Crinoids *Antedon* sp. and gooseneck barnacles on a dead sponge (crinoids 8 cm, depth 1,740 m, photo 5080-15 Young).

Below left:

Figure 129. Thalassometrid crinoid, flytrap anemone *Actinoscyphia* sp. 3, and brittle stars on the stem of a paramuricid gorgonian (anemone 12 cm, depth 785 m, photo 5093-110 Kelly).

Figure 130. Comatulid crinoid on sand-dusted hardpan (crinoid 15 cm, depth 1,620 m, photo 5087-138 Malahoff).

Below right:

Figure 131, *top right:* Unidentified crinoids on a limestone bench (crinoids 18 cm, depth 441 m, photo 5300-117 France).

Figure 132, *below right:* Stalked crinoid *Proisocrinus ruberrimus* on a basalt boulder near the submersible's collecting basket (crinoid 16 cm, depth 1,865 m, photo 5251-14 Bertram).

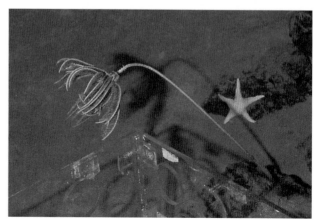

Crustaceans (Phylum Arthropoda)

Large crustaceans, such as shrimp and crabs, have hard carapaces of chitin covering the soft parts of their bodies; most have five pairs of jointed legs and stalked eyes. In the deep-sea areas around Hawaiʻi, barnacles, shrimp, crabs, and hermit crabs are fairly common; lobsters and mantis shrimp are fewer in number. Although many look like their shallow-water counterparts, some are very different.

Deep-dwelling barnacles are common in the Hawaiian region. When barnacles settle from the plankton, they turn upside down and cement themselves to the bottom. They then secrete hard protective plates around their bodies and use their feathery legs to rake in food particles from the water. Barnacles are a most varied and strange group of crustaceans. Some species live in cup-like shells cemented to rocky substrates. Others, like the gooseneck barnacle, are encased in plates attached to objects by a leathery stalk. The deep-sea gooseneck barnacles in figure 133 tended to be observed attached to dead cnidarian stalks and debris such as wood.

Unusual crab species include a very prickly animal that lives on extensive sand tracts far from ready hiding places. The carapace of this species is covered with long stout spines that probably protect it from the majority of predators (fig. 134). Another spiny crab is armed with longer, thinner spines. This crab also frequents sand flats, although it often perches on rocks (fig. 46) and crawls into crevices if disturbed.

An odd crab with a pair of long rake-like legs inhabits flat sandy or rocky bottoms. When approached, it holds both claws and rake-like

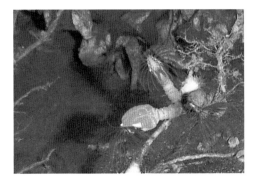

Figure 133. Group of barnacles *Alcockianum alcockianum* on a dead coral branch. A drowned coral reef is in the background (barnacles 8 cm, depth 1,640 m, photo 5064-50 Malahoff).

Figure 134. *Neolithodes* sp. on hardpan (crab 18 cm, depth 1,204 m, photo 5023-09 Malahoff).

Figure 135. Decorator crab *Paromala japonica* holding a glass sponge (*Sericolophus* sp.) on limestone and small pebbles (crab carapace 15 cm, depth 382 m, photo 281-13 Scheuer).

legs in front of its body, in what is probably a defensive posture (fig. 100). This species was observed eating such unappetizing creatures as glass sponges and sea cucumbers. Deep-sea decorator crabs use their fifth pair of legs to carry glass sponges and cnidarians over their carapaces (fig. 135). When the submersibles encountered one of these crabs, the animal turned around and displayed the animal so carried. Given that the particular sponges and cnidarians that the crabs were observed holding have especially large glass spicules or stinging cells, this type of behavior is most likely defensive (Wicksten 1985). Decorator crabs also live in flat sandy or rock areas and were often seen scuttling about rapidly. We clocked four crabs traveling at the rate of roughly 2 meters in 3 seconds.

Tiny galatheid crabs were seen only at depths below 200 meters, although they are known to occur in shallower water. They are abundant in most deep-sea environments and live in sponges (fig. 54), on cnidarians (fig. 63), in rock crevices (fig. 136), or in sand into which they can burrow rapidly.

Black-legged lobsters (fig. 137) live on limestone outcrops along the periphery of banks and benches that drop off abruptly into deep water. Large populations of this commercially valuable shellfish occur in limestone or basalt holes at depths of between 60 and 240 meters throughout the Hawaiian Archipelago and Johnston Atoll. Lobsters existing beyond scuba range, as these do, are an extremely valuable reserve for the future given the heavy exploitation of lobsters in shallow areas.

At 400 meters and deeper, the commercially important prawn *Heterocarpus laevigatus* inhabits crevices in eroded limestone near large sandy areas (fig. 136). This prawn and its smaller cousin *Heterocarpus ensifer* are fished using baited traps placed in the sandy areas. These prawns were seen moving from the rocks and onto the sand toward the traps shortly after they were set (Gooding et al. 1988).

A deep-sea banded prawn (fig. 138) is sometimes abundant in limestone holes on outcrops. Although these prawns live at depths to which very little light reaches, they are evidently able to see their surroundings since they swim from rock to rock instead of crawling on the bottom between outcrops. All underwater photographs of this large prawn show banded white and red individuals. When collected and brought to the surface, the animals lose their white bands. A species of red prawn has very long legs and antennae (fig. 139). Given that this animal lives in constant darkness, the length of its legs perhaps helps it locate both predators and prey. The ruby shrimp in figure 140 was observed alternately resting on and swimming above sand

Figure 136. The prawn *Heterocarpus laevigatus* and a galatheid crab *Munida heteracantha* on an eroded limestone reef (prawn 10 cm, depth 510 m, photo 5070-13 Moore).

Figure 137. Limestone ledge with lobsters *Panulirus marginatus* and a group of deep-reef fishes *Chaetodon miliaris*, *Chromis verater*, *Parupeneus pleurostigma*, and *Parupeneus porphyreus* (fishes 10–14 cm, depth 89 m, photo 382-51 Chave).

tracts at depths below 1,600 meters. This animal's antennae are even longer than those of the red prawn and trail along the bottom when the shrimp is skimming the sediment. If its antennae encounter a hard irregular bottom, the animal either backs up onto the sand again or swims upward.

One of the larger deep-sea hermit crabs is a scavenger and is often caught in baited shrimp traps. Its soft abdomen is covered by a brass-colored "shell" that is in turn covered by a large pinkish, stinging anemone (fig. 141). Juvenile hermit crabs begin their lives on the bottom looking for a mollusk shell in which to live. Next, the young hermit crab must find the right kind of anemone to place on its mollusk shell. The desired type of anemone secretes a protein that covers the shell and gives it its distinctive color. In time, the original shell dissolves or is discarded, and the crab lives in the brass-colored shell-like home maintained by the anemone (Dunn et al. 1980).

Figure 138, *top left: Plesionika pacifica* on an eroded limestone mound (prawn 10 cm, depth 349 m, photo 119-49 Chave).

Figure 139, *bottom left: Nematocarcinus tenuirostris* on a basalt tube (prawn 11 cm, depth 1,204 m, photo 5023-06 Malahoff).

Figure 140, *top right: Plesiopenaeus edwardsianus* over sand (shrimp 8 cm, depth 1,800 m, photo 5080-04 Young).

Figure 141, *bottom right: Parapagurus dofleini* and its anemone *Stylobates aenus* on a ropy lava flow (hermit crab 20 cm, depth 367 m, photo 116-30 McMurtry).

Sea Snails and Related Species (Phylum Mollusca)

Mollusks have a fleshy mantle, which may secrete a limy shell, and a creeping foot (divided into arms in the octopus and squid). Large-shelled species were seen only infrequently in deep-sea regions of the Hawaiian Ridge. Beds of brown pen shells (fig. 109) occur between depths of about 30 and 60 meters, on limestone flats unevenly covered with coarse sand. These mollusks are attached to hard substrates by tough threads connected to their shells, and, like most bivalves, they are filter feeders. An assortment of small fishes, invertebrates, and algae lives among these animals or attaches to their shells, and predators, like the knobby sea star and the helmet shell, live nearby.

One occasionally finds groups of collector shells on deeper sandy slopes; unfortunately, these animals are too small to photograph properly. Collector shells have the very interesting habit of obtaining coral fragments, worm tubes, and other mollusk shells and pasting them on top of their own shells as they grow. The final result is an amazing collage symmetrically arranged around the top of the shell. Although the reasons underlying this behavior are not well understood, perhaps the decoration camouflages them or strengthens their shells.

Shell-less mollusks are also seldom encountered in deep water. When they do appear, however, their antics are usually fascinating, and the submersible's pilots and observers take many photographs of them. The large buff nudibranch in figure 142 was studied by chemists who compared its biochemistry with that of the tan sponges on which it feeds. Octopuses and sometimes squid were occasionally seen on the bottom and in midwater. The large brown octopus (*Octopus* sp. B) in figure 143 is crawling on a sheet flow. This species has been caught, but more specimens must be collected before it can be determined whether it is a new species. The particular individual in the photograph squeezed into a small crevice when the submersible approached and disappeared.

Two species of tiny red octopuses occupy rocky outcrops. *Scaeurgus* has large eyes and smooth skin. It was photographed only once as it clung to the side of a limestone mound (fig. 144). *Berrya* has bumpy skin, and mantle tissue covers its eyes (fig. 145). It is an active animal, often seen crawling along on both smooth and rough substrates.

Figure 142. *Tritonia* sp. near a half-eaten tan sponge on limestone (nudibranch 8 cm, depth 367 m, photo 144-27 Scheuer).

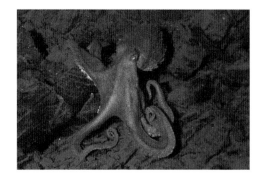

Figure 143. *Octopus* sp. B on a basalt tube (octopus 80 cm, depth 365 m, photo 116-26 McMurtry).

Figure 144. Black coral *Antipathes* sp. 1, small octopus *Scaeurgus patagiatus*, toadstool coral *Anthomastus fisheri*, and cidarid urchin *Stereocidaris hawaiiensis* on a limestone mound (urchin test 4 cm, depth 343 m, photo 119-90 Chave).

Variously sized individuals of an unknown *Octopus* species live on limestone substrates off Oʻahu and Molokaʻi. These animals were observed crawling about, poking their tentacles under rocks, sliding into cracks, or climbing onto or under different objects (fig. 146). Several pairs were seen mating during a series of dives at the Makapuʻu gorgonian beds in February and March 1983. Smaller individuals are more common than larger ones and often live in holes under small outcrops in sandy flats.

Several deep-sea octopuses spend most of their time swimming. One of these species, dubbed the eared octopus (fig. 147), was spotted swimming along in midwater; as the submersible approached, it moved to the bottom and coiled its short arms beneath its body. On a dive on Lōʻihi, a huge cirroteuthid octopus appeared when geologists were sampling rocks on the flanks of the volcano. It moved toward the submersible and spent several minutes "dancing" in front of the camera (fig. 148). It then extended its large webbed tentacles and jetted off, reappearing again in a few seconds. It was captured by the submersible's manipulator, and its head was maneuvered into the collecting basket. Unfortunately, as the submersible ascended, the octopus pulled away by grabbing a handle on the superstructure and escaped after inking the water.

Figure 145. Octopus *Berrya hoylei* on an eroded limestone mound (octopus 10 cm, depth 350 m, photo 217-09 Polovina).

Figure 146. *Octopus* sp. on eroded limestone (octopus 18 cm, depth 398 m, photo 148-24 Chave).

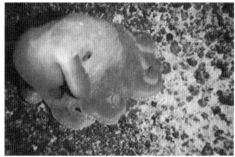

Figure 147. *Grimpoteuthis* sp. on poorly sorted sediment at Pensacola Seamount near Cross Seamount (octopus 8 cm, depth 1,576 m, video 5236 France).

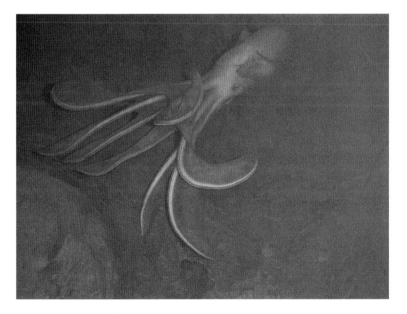

Figure 148. *Cirroteuthis* sp. in midwater with pillow basalts in the background (octopus 316 cm, depth 1,350 m, photo 5084-74 Malahoff).

Fishes (Phylum Chordata)

Sharks, rays, and chimaeras are primitive fishes with skeletons composed of cartilage. All are predators, regardless of whether they live in shallow or deep water.

Most deep-sea sharks are small and relatively similar in appearance. Many were photographed swimming away from the submersible and were therefore not easy to identify. The false cat shark is, however, an exception (fig. 149). Its body shape and markings are distinctive, and one of these animals allowed itself to be videotaped removing prawns from the mouth of a trap. The light from the submersible was very dim, yet the shark unerringly swam into the entrance of the trap several times, snapping at and eating the prawns there.

The smallest positively identified shark in the Hawaiian Islands is the gray dogfish (fig. 150). This species is usually seen resting on flat, smooth limestone. If disturbed, it swims for short distances and then settles back down to the bottom. The largest deepwater animals seen on the dives were six-gilled sharks (fig. 151). These fishes swam along the bottom in sandy areas without either approaching or veering away from the submersibles. One large six-gilled shark glanced off *Makaliʻi*'s hull, startling both pilot and observer. The shark seemed to be unaffected by this close encounter, moving away slowly and continuing to swim parallel to the submersible for a time. At the time of this writing, the submersibles have never encountered a great white or any of the other very large sharks known to inhabit the deep waters around the Hawaiian Archipelago.

Sometimes the submersibles attracted sharks. During several dives, especially in the Northwestern Hawaiian Islands and Johnston Atoll, groups of sharks followed the submersible as it descended to the bottom. The gray reef shark (fig. 152) was the most frequent of these visitors. Individuals or small groups often circled the submersible from the surface to depths of about 300 meters before veering away.

Five Cooke's sharks were resident on the summit of Cross Seamount and soon became a main attraction during *Pisces V* dives. Cooke's shark has an unusual body form, with most of its fins located near the tail (fig. 153). The sharks on Cross were very large, and each animal could be distinguished by its unique scars and markings. Individuals or pairs usually appeared about 10–20 minutes after the submersible reached the top of the seamount, slowly circling the area and disappearing a few minutes later. Occasionally they returned during the course of the dive.

Right:
Figure 149. False cat shark *Pseudotriakis microdon* near a trap on fine sediment (shark 80 cm, depth 500 m, video 5050 Polovina).

DEEP-SEA ANIMALS 59

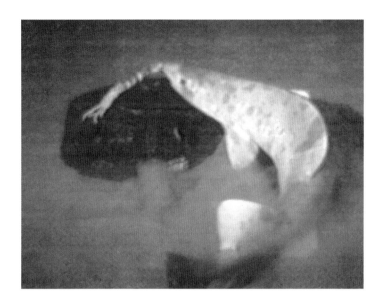

Figure 150, *middle left:* Gray dogfish *Squalus mitsukurii* over limestone (shark 40 cm, depth 398 m, photo 148-94 Chave).

Figure 151, *bottom left:* Six-gilled shark *Hexanchus griseus* over sand (shark 180 cm, depth 367 m, photo 43-31 Colin).

Figure 152, *top right:* Gray reef shark *Carcharhinus amblyrhynchos* near the limestone slope (shark 170 cm, depth 208 m, photo 195-58 Ralston).

Figure 153, *bottom right:* Cooke's shark *Echinorhinus cookei* swimming over sand (shark 200 cm, depth 410 m, photo 5039-43 Mullineaux).

Rays usually lie on or partially buried in sediment. The round ray inhabits sandy plains at depths greater than 300 meters. Its body is circular, often reaching 2 meters in width, and its tail is paddle-like (fig. 154). These rays cruise slowly over the bottom, searching for food buried in the sand. On detecting prey, they dig large holes to uncover it. Several deep holes were found during dives across sediment plains; sand clouds billowing upward indicated that one of these rays was foraging. The deeper, rarer, longnosed ray (fig. 155) has recently been discovered in the Indo-Pacific. Individuals were seen lying on sand or on smooth, hard surfaces at depths of between 750 and 1,000 meters off the island of Hawai'i. They can swim above the bottom for fairly long distances, but little else is known about them.

On one dive the submersible disturbed a small electric ray, which emerged briefly from the fine sediment (fig. 156). Organs in their disk-like bodies generate electric shocks that protect these rays against predators and stun prey. Fortunately for Hawai'i's beachgoers, these small rays do not live in shallow water.

Chimaeras are very strange, primitive, shark-like fishes. Rather than using their tails for swimming, they flap large, fan-like pectoral fins. Two species of chimaera have been seen in Hawaiian waters. The first, *Hydrolagus purpurescens,* has been encountered on several occasions on or near the bottom off Lō'ihi and the island of Hawai'i (fig. 157). The second, the longnosed chimaera, has only recently been observed and photographed off Lāna'i (fig. 158). Whereas *Hydrolagus* avoided the submersible, the longnosed species was attracted to it, swimming away only when the photo lights were turned on.

Figure 154. Round ray *Plesiobatis daviesi* on sand and unsorted dredge spoil material (ray 30 cm, depth 367 m, photo 87-18 Maragos).

Figure 155. Longnosed deepwater ray *Hexatrygon longirostra* on a pillow field with sand patches (ray 60 cm, depth 970 m, photo 5069-06 Moore).

Figure 157. Chimaera *Hydrolagus purpurescens* over coarse sediment and manganese crust talus (chimaera 25 cm, depth 1,220 m, photo 5238-11 France).

Figure 156. Electric ray *Torpedo* sp. over fine sediment (ray 12 cm, depth 352 m, photo 119-68 Chave).

Figure 158. Longnosed chimaera *Rhinochimaera pacifica* over fine sediment (chimaera 44 cm, 1,136 m, photo 5299-36 France).

Several bony fishes that mostly inhabit the open ocean have joined the submersibles from time to time. Often one or more black jacks (fig. 159) have accompanied the submersible on dives off Johnston Atoll, returning back toward the surface only after reaching depths of about 300 meters. Schools of rainbow runners (fig. 160) have also been observed as the submersibles descend.

It is the amberjack, however, that has been more commonly observed on dives. These large jacks followed *Makaliʻi* and *Pisces V* from the surface to depths of 350 meters, feeding on organisms scurrying away from the skids and manipulator or swimming in the light path. Wide, dark, oblique bands appear on their heads when they are feeding (fig. 161). Amberjack are attracted to the submersible's lights, motion, and noise. Occasionally one would ram into the submersible, breaking lights and dislodging specimens from collection baskets. These fishes had a most annoying habit of devouring the majority of specimens being collected by the submersible's manipulators.

The only other members of the jack family seen at depths greater than 300 meters are schools of deep-sea mackerels (fig. 162). If a school is present when the submersible's photo lights are turned on, the fishes rush toward the light, many individuals stunning themselves against the hull before the group moves away.

Eels are bony fishes that have adapted well to shallow- and deep-water benthic environments. Their snake-like bodies can easily slip into crevices both to avoid predators and to obtain prey, which is detected by extremely acute sensory organs. Many of the deep-sea eels caught by dredges and traps feed on shrimp, galatheid crabs, and prawns (Moffitt and Parrish 1992). The most abundant deep-sea species are cutthroat eels and conger eels. Cutthroat eels are so called because their mouths are very large and slit-like (fig. 163). Conger eels have smaller mouths. Both species are usually solitary, but, if bait is left on the bottom for a few minutes, all eels in the vicinity congregate (fig. 164).

The spotted snake eel is rare in Hawaiʻi but common at Johnston Atoll. During the day, snake eels aggregate under rocks; at night, they move about shallow, sandy reef flats. Because this eel has been reported to be a shallow-water species, it was surprising to find several individuals moving about the outer slopes of the Atoll at depths to 260 meters (fig. 165). In deeper water, large arrowtooth eels move across hard substrates near gold corals (fig. 166). Arrowtooth eels are often observed lying in the branches of these zoanthids, perhaps because prey is attracted to the bioluminescent light produced by the corals

Figure 159. Black jacks *Caranx lugubris* near the bottom (fish 40 cm, depth 122 m, photo 187-51 Ludwig).

Figure 161. Two amberjacks *Seriola dumerili* (fish 60 cm, depth 153 m, photo 96-06 Gooding).

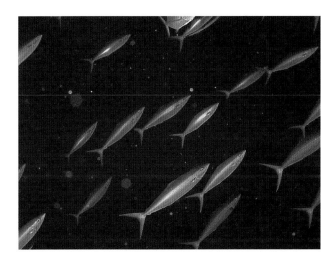

Figure 160. School of rainbow runners *Elagatis bipinnulata* in midwater (fish 25 cm, depth 122 m, photo 195-87 Ralston).

Figure 162. Mackerels *Decapterus tabl* over limestone (fish 15 cm, depth 404 m, photo 148-49 Chave).

Below left:

Figure 163. Cutthroat eel *Synaphobranchus brevidorsalis* swimming over manganese crust and a primnoid gorgonian (eel 50 cm, depth 1,260 m, photo 5237-48 France).

Figure 164. Small conger eels *Ariosoma marginatum* and prawns *Heterocarpus ensifer* around an opened tuna can placed on the sand (eels 15 cm, depth 400 m, photo 216-03 Polovina).

Below right:

Figure 165. Spotted snake eel *Myrichthys maculosus* on a steep sediment-dusted limestone slope dotted with pink coralline algae (eel 35 cm, depth 223 m, photo 202-17 Agegian).

Figure 166. Arrowtooth eels and a gold coral *Gerardia* sp. on limestone. A sea star *Plinthaster ceramoideus* is in the foreground (eels 60 cm, depth 1,200 m, photo 68-35 Chave).

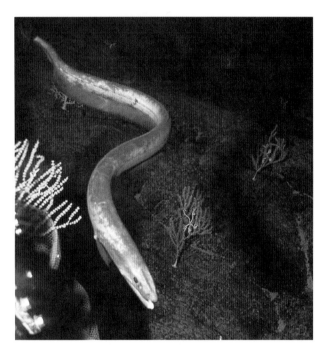

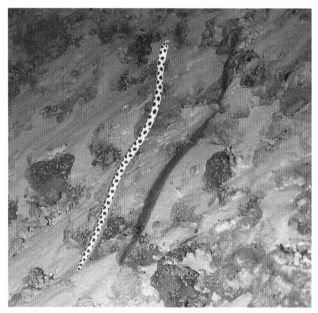

when they are touched. When currents are strong, eels aggregate on the lee side of the corals, using them for shelter.

Snappers range in depth from 10 to 400 meters and are considered to be some of the best-eating food fishes in Hawai'i. All the deepwater snappers described below are regularly fished along steep slopes.

Because there were no shallow-water snappers native to Hawai'i, the blue-line snapper was introduced from Tahiti in the 1950s. Since then, the species has spread throughout the Hawaiian Archipelago. Today, large aggregations of blue-line snappers frequent areas of high relief in relatively shallow water (fig. 167).

Small groups of pink snappers occur at depths below 50 meters. They usually swim about 1–5 meters above the bottom, where they feed on small invertebrates and fishes. Figure 168 depicts two of these fishes and some tangs visiting one of the artificial reefs placed on Penguin Bank by National Marine Fishery personnel.

Rosy snappers have the largest depth range of the Pacific snappers, occurring between 90 and 400 meters. These snappers are usually solitary (fig. 169) and during the day are equally likely to be found hovering near crevices or caves as venturing into midwater to swim along steep cliffs.

Below 150 meters one encounters banded snappers and ruby snappers. Banded snappers usually swim in pairs near cliffs, ruby snappers in small groups in midwater just above the bottom. Figure 170 shows

Figure 167. Aggregation of blue-line snappers *Lutjanus kasmira* around an outfall (fish 16 cm, depth 30 m, photo 143-12 Russo).

Below left:

Figure 168. Two pink snappers *Pristipomoides filamentosus* (center, right) and tangs *Naso hexacanthus* and *Naso maculatus* near an artificial reef (snapper 30 cm, depth 118 m, photo 374-06 Moffitt).

Figure 169. Rosy snapper *Etelis carbunculus* over limestone and debris (fish 30 cm, depth 398 m, photo 213-74 Chave).

Below right:

Figure 170. Banded snapper *Pristipomoides zonatus* in a small limestone cave (snapper 20 cm, depth 193 m, photo 195-35 Ralston).

Figure 171. Ruby snapper *Etelis coruscans* over pillow lava flow (fish 30 cm, depth 352 m, photo 231-161 McMurtry).

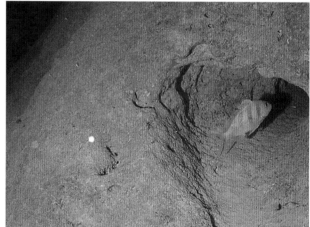

a young banded snapper near a hole in a limestone cliff, figure 171 a ruby snapper. Ruby snappers are bright red and have long tail fin rays. Unlike their close relatives the rosy snappers, they seem never to rest on or hover over the bottom.

Figure 137 shows some of the deeper reef fishes that live at depths of 20–100 meters. During the day these animals aggregate in areas of high relief along the margins of banks or on ledges above drop-offs. The goatfishes move into sandy areas at night and probe for prey with their barbels. Most damselfishes and butterflyfishes are, however, quiescent at night, having spent the daylight hours feeding on plankton or picking items from hard substrates. Figure 79 shows a group of soldierfishes and squirrelfishes in a cavern. These two fish species inhabit holes and caves during the day and feed in midwater at night.

There are very few large groupers in shallow or deep Hawaiian waters, and only the Hawaiian grouper is native. This species lives in a variety of habitats, depending on the age of the fish. Young animals frequent holes and caves at depths as shallow as 15 meters, and adults live in similar habitats at depths of between 45 and 315 meters. Adults appear to be territorial. At Johnston Atoll, identifiable Hawaiian groupers were seen in the same places on repetitive dives. On dives at Johnston Atoll and Maui, they were seen chasing other large predators away from some of the holes and crevices in limestone ledges.

Hawaiian groupers often look into the submersible's ports or follow the submersible, especially if the manipulator has picked up objects (fig. 172). On one dive, a grouper swam into the submersible's basket and began to eat the collected animals—to the great annoyance of the pilot, who had taken great care to collect appropriate specimens.

The red and white grouper (fig. 173) lives at depths of over 250 meters. These tiny fishes dart about the ledges, picking at objects on rocks and in sand patches. They hid in crevices when the submersible's bright photo lights were turned on.

Two attractive basslet species inhabit midwater near rocky caves at depths of roughly 50–300 meters. In the submersible's photo lights, the darkfin basslet appears yellow and Elizabeth's basslet white and brown (fig. 174). In close-up photographs, however, or when brought to the surface, the darkfin basslet is actually orange and lavender and Elizabeth's basslet white, orange, and lavender (a photograph of a darkfin basslet appears in Severns and Fiene-Severns 1993). Darkfin

Figure 172. Hawaiian grouper *Epinephelus quernus* on limestone (fish 50 cm, depth 260 m, photo 125-21 Devaney).

Figure 173. Red and white grouper *Plectranthias kelloggi* and a scorpionfish on an eroded limestone bench (fish 6 cm, depth 260 m, photo 125-12 Devaney).

Figure 174. Male basslet *Holanthias elizabethae* near a lava tube (fish 8 cm, depth 199 m, photo 291-18 Moore).

basslet males, females, and juveniles occur together in small groups; Elizabeth's basslets occur in pairs (Chave and Mundy 1994).

Small groups of deep-sea butterflyfishes (fig. 37) frequent slopes near holes and caves at depths of between 100 and 300 meters. These fishes always occur close to the bottom and have never been seen picking at the bottom or nipping at objects in the water column. A videotape taken on a dive off Moloka'i shows a butterflyfish "cleaning" larger fishes; perhaps at least part of the diet of deep-sea butterflyfishes consists of ectoparasites. (Scientists in submersibles have, in fact, observed several instances of these and other deep-sea fishes being cleaned.) However, our understanding of their dietary habits will remain incomplete until more of these rarely captured animals are taken.

A colorful deepwater bigeye (fig. 175) frequents island slopes at depths of 130–300 meters. Because it is attractive and slow moving, this fish has been a favorite photographic subject of submersible observers and pilots. Adults vary in color from silvery to dark red and may darken or fade within seconds. Individuals, pairs, or small groups of these animals hover or move slowly over hard substrates near caves or ledges. Young animals are usually mottled and live in sandy depressions formed by currents swirling over small outcrops or debris.

Figure 175. Two bigeyes *Cookeolus japonicus* on an eroded limestone slope (fish 18 cm, depth 214 m, photo 97-39 Gooding).

Figure 176. Aggregation of small pink snappers *Symphysanodon typus* and yellowfin soldierfish *Myripristes chryseres* near a limestone cave (snappers 7 cm, depth 153 m, photo 337-48 Ralston).

Two species of small red snapper-like fishes form large aggregations on deep slopes at depths of between 80 and 400 meters (fig. 176). These are the most abundant of the fishes in this depth range, but they are too small to be of commercial value. They dart about in midwater, nipping at objects; when approached, they swim toward the bottom, hiding in caves or holes.

Several groups of fishes with large eyes inhabit the twilight zone at depths between 200 and 500 meters. The scorpionfishes found there resemble their shallow-water counterparts. Stout poisonous spines usually protect scorpionfishes themselves from predators, so they generally remain motionless until they detect prey. Apparently they are able to see their prey because, when edible objects pass by, the fishes snap at them with their large mouths.

Two of the deepwater scorpionfishes that inhabit rocky areas near holes and crevices can be seen in figures 18 and 177. Young Guenther's scorpionfish are brownish with yellow borders on their fins. For unknown reasons, they become brick red as they mature (fig. 18). The largeheaded scorpionfish (fig. 177) has extremely long tentacles above its eyes that may help it detect prey.

Many fishes inhabiting the twilight zone belong to groups not seen in shallow water. Individuals or small groups of boarfishes hover in areas of hard, irregular terrain (fig. 178). Like the deep-sea butterflyfishes, they were observed nudging, perhaps cleaning, parasites from larger fishes.

A newly discovered species of bandfish and the goldspot flathead live at the bases of outcrops. Bandfishes are red and have long pointed tails; males also have long white pelvic fins (fig. 179). Males and females usually live separately in or near holes on limestone flats. These fishes are extremely difficult to capture because at the slightest disturbance they dive into crevices.

The goldspot flathead, a bluish fish with gold spots and a large mouth, is usually observed lying on the substrate; occasionally it can be seen darting across the substrate or into the water column to catch prey. It is possible that it hunts by sight alone, but other sensory modes that detect water disturbances or odors may be involved.

Figure 177. Largeheaded scorpionfish *Pontinus macrocephalus* on a limestone bench (fish 25 cm, depth 260 m, photo 125-04 Devaney).

Figure 178. Boarfish *Antigonia* sp. over a sand patch near a lava flow (fish 9 cm, depth 239 m, photo 174-56 Chave).

Figure 179. Pink bandfish *Owstonia* sp. and flathead *Chrionema squamiceps* on eroded limestone (*Chrionema* 9 cm, depth 367 m, photo 215-22 Chave).

The white-spotted spikefish was so named by the submersible crew because of the white spots dotting its pinkish red body (fig. 180). Spikefishes have large dorsal fin spines, probably used for protection, and strong ventral spines. These fishes occupy hard flat substrates near outcrops, often resting on the bottom, perched on their ventral fin spines.

The spotted lanternbelly was the most quiescent of all fishes encountered by the submersible (fig. 181). During the day it lay on outcrops and usually did not move, even when poked by the manipulator. This species is bioluminescent and actively pursues its prey—shrimp and other fish—in the water column at night (Reid et al. 1991).

Deep-sea spotted pufferfish are sometimes seen in sandy areas where their prey—squid and prawns—is plentiful. The pufferfish in figure 182 was photographed eating prawns from one of the experimental traps set on the Mauna Kea Ledge.

Often encountered in the videotapes is a large-mouthed flatfish (fig. 183). Flatfishes are able to change their body coloration to match the substrate on which they are resting. Therefore, observers usually noticed only those animals that moved when they were disturbed by the submersible.

Red rakefish live at depths below 200 meters on or near limestone benches (fig. 33). Rakefishes have enormous eyes, distinctive rake-like barbels, and bumper-like processes on the front of their heads. Their barbels have taste organs at the tips that are used to detect food, and the fish use their bumpers to dig out prey so detected. Two finger-like rays on each pectoral fin are used for walking along the bottom. If disturbed, rakefishes swim for short distances. The orange rakefish (fig. 184) was occasionally seen walking along sandy areas. This little animal has shorter bumpers on its head than the red rakefish.

Alfonsin live at depths between 400 and 500 meters along the tops of current-swept ledges near drop-offs (fig. 185). They are important food fishes found in abundance on several seamounts in the Northwestern Hawaiian Islands. HURL bottom-camera photographs show these red fishes aggregating near the edge of Cross Seamount during the day and over its top at night. They eat midwater shrimp.

The john dory (fig. 186) and the blue snake mackerel (fig. 187) are important food fishes in the temperate western Pacific. Both species are spectacular animals, and the submersible crew liked to photograph and videotape them. However, the john dorys encountered always seemed to be out of camera reach, and blue gempylids move about very swiftly. As a consequence, more bad photographs were taken of these two animals than of most other popular fishes.

Figure 180. White-spotted spikefish *Hollardia goslinei* near a limestone mound (fish 6 cm, depth 346 m, photo 118-35 Chave).

Figure 181. Spotted lanternbelly *Synagrops argyrea* on a limestone mound (fish 14 cm, depth 346 m, photo 118-18 Chave).

Figure 182. Deep-sea spotted pufferfish *Sphoeroides pachygaster* on sand (fish 19 cm, depth 367 m, photo 219-31 Gooding).

DEEP-SEA ANIMALS 71

Figure 183. Flatfish *Chascanopsetta prorigera* on pitted limestone (fish 15 cm, depth 352 m, photo 107-25 Chave).

Figure 184, *middle left:* Orange rakefish *Satyrichthys hians* on coarse sand (fish 7 cm, depth 550 m, video 5174 Kerby).

Figure 185, *bottom left:* Small alfonsin *Beryx decadactylus* over limestone off Hawai'i (fish 20 cm, depth 367 m, photo 215-05 Chave).

Figure 186, *top right:* John dory *Zenopsis nebulosus* and the manipulator collecting whip bamboo coral over limestone (fish 25 cm, depth 367 m, photo 357-37 Grigg).

Figure 187, *bottom right:* Blue snake mackerel *Rexea nakamuri* on eroded limestone (fish 10 cm, depth 398 m, photo 213-61 Chave).

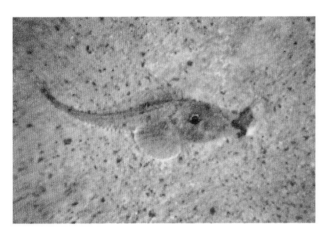

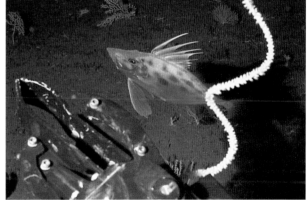

Eel-like fishes are mostly found in sandy areas at depths of over 400 meters. These animals react to the submersible in different ways. Halosaurid fishes frequent channels between the islands (fig. 188). Light, noise, and water movement attract them, and individuals followed *Pisces V* for periods of up to 15 minutes. The jellynose eel (fig. 189), so named because of its bulbous jelly-like snout, frequents extensive sand flats. Individuals that the submersible came on became completely disoriented in its running lights, and, when the bright photo lights were turned on, they froze, drifting with the current until the lights were extinguished. Thick-bodied cusk eels (fig. 190) and slender duckbill eels (fig. 191), found on sand patches near basalt or limestone outcrops, moved into crevices when approached by the submersible.

Spiderfishes and tripodfishes are perhaps the strangest fishes encountered by *Pisces V*. These animals live at depths below 1,100

Figure 188, *top left:* Halosaurid *Aldrovandia phalacra* over sand and pebbles (fish 40 cm, depth 1,360 m, photo 5032-29 Jones).

Figure 189, *bottom left:* Jellynose eel *Ijimaia plicatellus* over coarse sediment (fish 40 cm, depth 360 m, photo 34-146 Colin).

Figure 190, *top right:* Cusk eel *Pycnocraspedum armatum* in a sand-filled crevice between basalt flows (eel 16 cm, depth 750 m, photo 5068-49, Moore).

Figure 191, *bottom right:* Duckbill eel *Nettastoma parviceps* in a sand channel between basalt flows (eel 25 cm, depth 780 m, photo 5068-11 Moore).

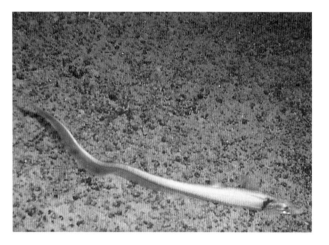

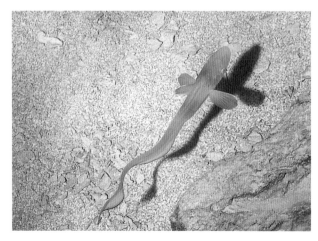

meters on vast sand flats. Spiderfishes have elongated, whisker-like pectoral fin rays (fig. 192), tripodfishes very long pelvic and lower caudal fins (fig. 193). Both species rest motionless on their fins, their bodies held above the sediment, a posture that most likely prevents abrasion by sand particles. If disturbed, they swim normally for a short distance, their long fins trailing behind them, and then make perfect three-point landings on the sand, always facing the current. The scientists who observed spiderfishes and tripodfishes from the submersible reported that these fishes occasionally snapped at small objects brushing against them.

A remote relative of the spiderfish and the tripodfish, the Hawaiian greeneye, dots sediment plains at depths between 100 and 400 meters. Its fusiform body has a metallic sheen, and its eyes reflect red in photographs (fig. 194). During the day, greeneyes lie on the bottom,

Figure 192, *top left:* Spiderfish *Bathypterois atricolor* on fine sediment (fish 10 cm, depth 1,800 m, photo 5080-03 Young).

Figure 193, *bottom left:* Tripodfish *Bathypterois grallator* on rippled sediment (fish 11 cm, depth 1,290 m, photo 5062-58 Malahoff).

Figure 194, *top right:* Two greeneyes *Chlorophthalmus proridens* on a fine sediment plain (fish 8 cm, depth 367 m, photo 123-13 Chave).

occasionally darting into the water column to catch small, moving prey. Large numbers of Hawaiian greeneyes have been caught in trawls at night with their stomachs full of midwater shrimp. It appears that these fishes feed both day and night.

Some of the Hawaiian deep-sea fishes that live at depths between 400 and 2,000 meters have been observed using sensory filaments to detect prey (Chave and Mundy 1994). Spiderfishes, for example, snap when their whisker-like appendages encounter objects on the sand.

Grenadiers (rattails) have a sensory barbel located under the chin. Two members of this species were observed feeding. Observations indicate that the spotted rattail usually swims slowly over or rests on sand (fig. 195), occasionally also hovering above the sediment in a head-down position, waiting until its barbel detects prey in the sand. When food is located, it roots out the item with its undercut jaw and triangular armored snout. The black grenadier (fig. 196) employs the same feeding technique but spends more time swimming above the bottom.

Cods use extended sensory rays on their pelvic fins to locate food. The black cod seen in figure 33 occupies holes in the limestone and probes for food with white, filamentous pelvic fins.

Beardfishes (fig. 197) often travel in groups and are very active swimmers. They probe the sand with two flexible white barbels and, when small shrimp or other food items are detected, snap them off the bottom. The submersible's movements and lights often attracted beardfishes. They were observed crossing and recrossing the light paths or approaching the manipulators, and one animal was videotaped sticking its snout in a coring device.

Hawaiian species of deep-sea anglerfishes live at depths between 300 and 2,000 meters. The tan goosefish has been seen angling for its dinner with two luminous lures, the gray goosefish using three lures to attract and eat two mackerels.

Tan goosefishes are globular in shape (fig. 198). They usually perch motionless on rocks, although they occasionally creep from one perch to another on their hand-like fins and are also able to swim. Videotape footage reveals that capturing goosefishes with the submersible's manipulators can be a challenge.

The gray goosefish has a flattened body, a large head, and an enormous mouth (fig. 199). Fringe-like sensory flaps run along the middle of its body at the border between top and bottom, and its luminous lures are modified parts of its dorsal fin. This species of goosefish has been observed doing "push-ups" on its frog-like pectoral fins.

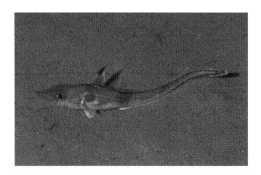

Figure 195. Spotted rattail *Caelorinchus spilonotus* on fine sediment off Hawai'i (fish 15 cm, depth 367 m, photo 123-20 Chave).

Figure 196. Black grenadier *Coryphaenoides longicirrhus* skimming over the sand (fish 454 cm, depth 1,450 m, video 5180 Malahoff).

Figure 197. Beardfish *Polymixia berndti* over limestone and debris off O'ahu (fish 20 cm, depth 373 m, photo 213-48 Chave).

Sea toads differ in color from the tan goosefish, but their body shape and habits are similar (figs. 1 and 200). They are usually to be found perched on rocks and occasionally waddle slowly along on the bottom, swimming only when prodded. Like goosefishes, sea toads have enormous mouths and expandable bodies adapted for swallowing very large prey. When the fringe-like flaps on their bodies sense movements in the water, they flick up a rod-like dorsal fin spine. The luminescent lure on the spine attracts small swimming animals, which the sea toad promptly gulps down.

Figure 198. Tan goosefish *Sladenia remiger* perched on a pillow (fish 25 cm, depth 1,270 m, photo 5186-141 Garcia).

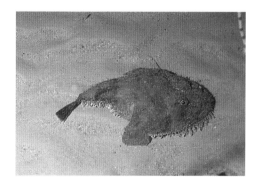

Figure 199. Gray goosefish *Lophiodes miacanthus* on sand-dusted limestone (fish 35 cm, depth 367 m, photo 144-17, Scheuer).

Figure 200. Orange sea toad *Chaunax fimbriatus* on sediment and basalt (fish 9 cm, depth 1,985 m, video 5192 Bertram).

Table 1
A Taxonomic List of Animals Photographed from HURL Submersibles

TAXON	Rank* (observed)	Depth (m) (observed)	Substrate (on/above)	Location** (observed)	Depth (m) (published)	Location (published)	Selected References (depth, location, and/or photograph)
Phylum Porifera							
Class Hexactinellida							Gray 1867, 1870
Hexactinellid holey	R	1,430–1,810	Talus	LH			Gray 1867, 1870; Schultze 1899; Ijima 1901; Tabachnick 1991
Hexactinellid spiny	R	920–970	Limestone	LA, PB			
Hexactinellid ribbon	O	150–920	Hard	HI			
Hexactinellid stalked	R	1,205–1,550	Limestone	PB			
Hexactinellid tan	A	61–1,950	Limestone	HI			
Hexactinellid waffle	R	310–640	Hard	HI			
Family Euplectellidae							
Bolosoma sp.	C	920–2,050	Hard	CS, LH, LV			H. Reiswig, pers. comm., 1997
Corbitellinae	O	1,180–1,725	Hard	CS, LH			H. Reiswig, pers. comm., 1997
Corbitella sp.	R	1,340–1,865	Basalt	LA, MO, PB			Ijima 1902
Dictyaulus sp.	R	790–1,940	Basalt	CS, LH			H. Reiswig, pers. comm., 1997
Regadrella sp. 1	O	318–413	Hard	HA			
Regadrella sp. 2	R	380–1,367	Hard	CS, WO			H. Reiswig, pers. comm., 1997
Regadrella sp. 3	R	500–1,448	Basalt	CS, LH			H. Reiswig, pers. comm., 1997
Sericolophus sp.	O	355–550	Sand	LH, LO, WO			
Trachicaulus sp.	R	1,515–1,940	Basalt	LV			
Walteria flemmingi Schultze	O	940–1,500	Basalt	CS, LO	Unknown	Unknown	Schultze 1886; H. Reiswig, pers. comm., 1997
Walteria cf. *leuckarti* Ijima	O	605–2,040	Basalt	CS, HI, LV	1,570–1,660	Pacific	Ijima 1901, Schultze 1887
Walteria sp.	O	1,225–1,840	Hard	CS, LH, LV			
Family Lanuginellidae							
Crateromorpha sp.	R	1,300–1,945	Hard	LH, PB			
Family Euretidae							
Chonelasma sp.	R	780–1,770	Hard	LA, LH			

(continued on next page)

* NUMBER OF PHOTOGRAPHED ANIMALS RANKED AS FOLLOWS: A (abundant), greater than 100; C (common), 50–99; O (occasional), 10–49; R (rare), 1–9.

** LOCATIONS ARE CODED AS FOLLOWS: CS, Cross Seamount; FFS, French Frigate Shoals; HA, Hawaiian Archipelago; HI, high Hawaiian Islands; JA, Johnston Atoll; LA, Lāna'i; LH, leeward Hawai'i; LO, leeward O'ahu; LV, Lō'ihi Volcano; M, windward Maui; MO, Moloka'i; NWHI, Northwestern Hawaiian Islands; PB, Penguin Bank; WH, windward Hawai'i; WO, windward O'ahu.

Table 1
A Taxonomic List of Animals Photographed from HURL Submersibles (continued)

TAXON	Rank* (observed)	Depth (m) (observed)	Substrate (on/above)	Location** (observed)	Depth (m) (published)	Location (published)	Selected References (depth, location, and/or photograph)
Family Farreidae							
Farrea occa Carter	O	580–2,012	Hard	LH	1,815–1,980	Indo-Pacific	Tabachnick 1988
Farrea sp.	R	710–1,985	Basalt	LH			
Family Pheronematidae							
Semperella sp. 1	O	1,610–1,740	Limestone	LH			H. Reiswig, pers. comm., 1997
Semperella sp. 2	O	850–1,210	Limestone	LH			
Family Hyalonematidae							
Cyliconema sp. 1	O	945–1,950	Hard	CS, LH			
Cyliconema sp. 2	R	1,640–2,102	Hard	CS, LH, LV			
Hyalonema sp.							
Family Caulophacidae							
Caulophacus sp. 3	R	1,550–1,975	Hard	CS, LH, PB			
Family Dactylocalicidae							
Vase-like	R	1,130–1,230	Basalt	CS, LH			
Class Demospongiae							
Brown encrusting	O	128–416	Hard	LO, PB			
Cone-shaped	O	49–110	Hard	HI			
Cup-shaped	O	128–416	Basalt	HI			
Orange, red, rust, salmon, yellow, encrusting	A	31–370	Hard	HA, JA			
Round	A	49–1,950	Hard	HI			
Tube-like	A	49–1,865	Basalt	LH, WH			
Family Corallistidae							
Corallistes sp.	R	346–400	Basalt	HA			deLaubenfels 1963
Family Stylochordylidae							
Stylocordyla boreale (Loven)	R	1,350–1,370	Basalt, coarse sediment	LH	13–2,950	Circumglobal	Hartman 1982
Phylum Brachiopoda							
Terebratula sp.	R	1,950–1,975	Basalt	LA, PB			Davidson 1880, Foster 1982
Phylum Ctenophora							
Lyrocteis sp.	O	306–376	Hard	HA, JA			Robilliard and Dayton 1972
Phylum Pogonophora							
Sclerolinium sp.	R	1,040–1,045	Vent	LV			E. C. Southward, pers. comm., 1992
Phylum Annelida							
Potamilla sp.	R	367–376	Sand	LO			Bailey-Brock 1972

Phylum Bryozoa						
Bryozoan fan-shaped	A	55–391	Hard	HI	Hawaiian Is.	Soule et al. 1986
Family Smittinidae						
Parasmittina delicatula (Brusk)	C	52–192	Limestone, coarse sediment	LO, PB, WO	Shallow	Soule et al. 1986
Family Seretellidae						
Reteporellina denticulata (Brusk)	O	57–153	Limestone, coarse sediment	LO, PB, WO	1–100	Indo-Pacific	Soule et al. 1986
Family Schizoporellidae						
Schizoporella crassomuralis (Canu and Bassler)	A	49–116	Limestone, coarse sediment	LO, PB, WO	35–50	Indo-Pacific	Soule et al. 1986
Family Steginoporellidae						
Steginoporella magnilabris (Brusk)	A	52–161	Limestone, coarse sediment	LO, PB, WO	Shallow	Circumglobal	Soule et al. 1986
Phylum Cnidaria						
Order Hydroida						
Hydroid	C	83–1,740	Hard	HA, JA			
Hydrodendrium gorgonoides Nutting	O	336–450	Hard	LH, LO, WH	365–541	Hawaiian Is.	Nutting 1905
Order Milleporina						
Millepora tenera (Boschma)	O	37–70	Limestone	JA	1–70	Indo-Pacific	Maragos and Jokiel 1986
Order Stylasterina						
Distichopora violacea (Pallas)	C	40–229	Limestone	JA	1–206	Indo-Pacific	Maragos and Jokiel 1986
Stylaster sp.	O	228–401	Limestone	JA			
Order Alcyonacea						
Anthomastus fisheri Bayer	C	275–1,325	Hard	CS, HI	356–462	Hawaiian Is.	Bayer 1952
Anthomastus granulosus Kukenthal	R	73–367	Hard	LO, PB, WO	20–201	Hawaiian Is.	Bayer 1952
Anthomastus sp.	R	1,205–2,050	Basalt	NWHI			
Anthomastus steenstrupi Wright and Studer	R	975–1,060	Hard	LH, MO	195–241	Hawaiian Is.	Wright and Studer 1889, Bayer 1952
Sinularia abrupta Tixier-Durivault	R	38–40	Limestone	JA	1,031–1,033	Pacific	Bayer 1952
Nepthied sp. 1	R	920–1,000	Hard	CS, LH			
Nepthied sp. 2	R	344–367	Limestone cave	JA			
Order Stolonifera						
Clavularia sp. (purple)	C	349–416	Limestone	WO			
Clavularia sp. (yellow)	O	361–910	Basalt	CS, LH			
Order Telestacea						
Telestula spiculicola (Nutting)	O	61–183	Hard	HA	409–616	Hawaiian Is.	Nutting 1908, Bayer 1952

(continued on next page)

Table 1
A Taxonomic List of Animals Photographed from HURL Submersibles (continued)

TAXON	Rank* (observed)	Depth (m) (observed)	Substrate (on/above)	Location** (observed)	Depth (m) (published)	Location (published)	Selected References (depth, location, and/or photograph)
Order Gorgonacea							
Gorgonian broom	R	255–260	Limestone	PB			
Gorgonian brush	R	1,055–1,060	Hard	CS			
Gorgonian gold	R	1,140–1,370	Hard	CS			
Gorgonian gray	R	356–382	Limestone	LH			
Gorgonian leather	O	153–309	Hard	HI			
Gorgonian pink fan	R	1,550–1,575	Basalt	LH, LV			
Gorgonian purple	O	1,010–1,448	Hard	CS			
Gorgonian red	R	1,280–2,012	Basalt	LV, MO			
Gorgonian spider	R	1,290–1,652	Hard	LH			
Gorgonian yellow	R	1,000–1,010	Hard	CS			
Family Anthothelidae							
Anthothelia nuttingi Bayer	R	347–382	Hard	CS, LH	340–1,820	Hawaiian Is.	Bayer 1956, Grigg and Bayer 1976, France et al. 1996
Family Corallidae							
Corallium ducale Bayer	R	1,316–1,380	Basalt	CS	1,000–1,460	East Pacific	Bayer 1956, France et al. 1996
Corallium regale Bayer	O	229–1,315	Hard	CS, HA	365–719	Central Pacific	Bayer 1956, Grigg and Bayer 1976, Tabachnick 1981
Corallium secundum Dana	A	183–1,380	Hard	CS, HA	275–564	Hawaiian Is.	Bayer 1956, Grigg and Bayer 1976
Corallium tortuosum Bayer	R	245–370	Limestone	NWHI	167–408	Hawaiian Is.	Bayer 1956, Grigg and Bayer 1976
Corallium sp. (yellow)	C	558–1,496	Limestone	LA, LO, MO			S. C. France, pers. comm.
Family Briareidae							
Paragorgia dendroides Bayer	O	490–1,910	Hard, sand	CS, HI, LV	774–1,800	Pacific	Bayer 1956, Tabachnick 1981
Paragorgia regalis Nutting	C	332–1,190	Hard	CS, WO	350–396	Pacific	Kukenthal 1924, Grigg and Bayer 1976
Paragorgia sp.	R	367–785	Hard	CS, WO			
Family Acanthogorgidae							
Acanthogorgia sp.	R	1,290	Hard	CS			F. M. Bayer, pers. comm., 1996
Acanthogorgia striata Nutting	C	338–413	Limestone	WO	80–564	Pacific	Grigg and Bayer 1976
Family Melithaeidae							
Acabaria bicolor (Nutting)	O	122–260	Limestone	HI	73–408	Hawaiian Is.	Bayer 1956, Devaney 1976
Family Paramuriceidae							
Bebryce brunnea (Nutting)	C	339–575	Limestone	WO	168–387	Hawaiian Is.	Grigg and Bayer 1976
Paramuricid blue	A	180–1,530	Hard	CS, HI			
Paramuricid tan	A	251–1,500	Hard	CS, HA			
Villogorgia sp.	R	336–352	Limestone	PB			

Taxon	Abundance	Range 1	Substrate	Codes	Range 2	Region	References
Family Chrysogorgiidae							
Chrysogorgia geniculata (Wright and Studer)	O	1,010–1,925	Limestone	CS, HI, LV	148–1,135	Pacific	Nutting 1908, Kukenthal 1924
Chrysogorgia scintillans Bayer and Stefani	O	580–2,050	Basalt	CS, HI, LV	1,759–1,936	Pacific	Bayer and Stefani 1988
Chrysogorgia stellata Nutting	O	710–2,050	Basalt	LH, LV, MO	649–922	Pacific	Grigg and Bayer 1976, Bayer and Stefani 1988
Chrysogorgiid feathery	O	1,595–2,012	Basalt	LV			
Iridogorgia bella Nutting	R	750–1,925	Hard	CS, LH, LV	742–1,005	Hawaiian Is.	Nutting 1908, Kukenthal 1924
Iridogorgia superba Nutting	O	540–1,960	Hard	CS, LH, LV	704–914	Hawaiian Is.	Nutting 1908, Kukenthal 1924
Metallogorgia melanotrichos (Wright and Studer)	O	251–2,050	Hard	CS, HI, LV	183–1,500	Pacific	Nutting 1908, Grigg and Bayer 1976, Tabachnick 1981
Pleurogorgia sp.	R	1,335–2,050	Basalt	CS			
Family Ellisellidae							
Ellisella sp.	C	300–2,050	Basalt	CS, LH, LV			
Family Primnoidae							
Callogorgia gilberti Nutting	C	339–490	Limestone	CS, HI	215–960	Hawaiian Is.	Nutting 1908, Grigg and Bayer 1976
Callogorgia sp. 1	R	352–391	Limestone	WO			
Callogorgia sp. 2	C	332–407	Limestone	WO			
Calyptrophora agassizii Studer	C	336–435	Hard	LH, WH, WO	691–1,145	Pacific	Kukenthal 1924, Grigg and Bayer 1976
Calyptrophora clarki Bayer	O	214–321	Limestone	HA	12–1,275	Indo-Pacific	Grigg and Bayer 1976
Calyptrophora japonica Gray	R	311–413	Limestone	WO	216–1,300	Pacific	Grigg and Bayer 1976, Tabachnick 1981, Ohta 1983
Calyptrophora sp. 3	R	267–398	Limestone	WO			
Calyptrophora sp. 4	R	1,235–1,380	Limestone	LH			
Calyptrophora wyvilli Wright	C	975–1,370	Basalt	CS	744–823	Hawaiian Is.	Grigg and Bayer 1976
Candidella helminthopora Nutting	C	935–1,755	Basalt	CS, LH, MO	38–1,820	Hawaiian Is.	Nutting 1908, Grigg and Bayer 1976
Narella bowersi Nutting	O	800–1,448	Basalt	CS, HI	700–1,936	Hawaiian Is.	Nutting 1908, Grigg et al. 1987
Narella megalepis Kinoshita	C	260–398	Limestone	HA	215–564	Pacific	Grigg and Bayer 1976
Narella sp. 1	A	327–515	Limestone	LO, WH, WO	353–417	Hawaiian Is.	Grigg and Bayer 1976
Narella sp. 2	R	336–367	Limestone	WO			
Narella sp. 3	O	885–1,950	Limestone	CS, LH, LV			
Narella sp. 4	O	1,200–1,380	Basalt	CS			
Narella sp. 5	O	1,170–1,370	Basalt	CS, HI			
Paracalyptrophora angularis (Nutting)	R	875	Limestone	PB	unknown	Hawaiian Is.	S. C. France, pers. comm., 1996
Plumarella sp.	R	246–367	Limestone	M, WO			Nutting 1908; Bayer, pers. comm., 1996
Primnoid comb	O	1,806–1,954	Limestone	MO			
Primnoid harp	O	830–1,865	Hard	CS, LH, PB			

(continued on next page)

Table 1
A Taxonomic List of Animals Photographed from HURL Submersibles (continued)

TAXON	Rank* (observed)	Depth (m) (observed)	Substrate (on/above)	Location** (observed)	Depth (m) (published)	Location (published)	Selected References (depth, location, and/or photograph)
Primnoid lyre	R	780–1,900	Hard	CS			
Thouarella hilgendorfi Versluys	C	367–416	Limestone	WO	173–396	Pacific	Kukenthal 1924, Grigg and Bayer 1976, Bayer 1981
Family Plexauridae							
Eunicella sp. 1	A	332–450	Limestone	LO, WH, WO			
Euplexaura sp.	O	291–355	Limestone	JA			
Family Isididae							
Acanella sp. 1	C	332–630	Various	CS, HI	215–565	Hawaiian Is.	Grigg and Bayer 1976
Acanella sp. 2	O	355–1,925	Basalt	CS, HI, LV			
Acanella sp. 3	R	1,500	Hard	CS			
Isidella sp. 1	O	1,500–1,630	Basalt	CS			France et al. 1996
Isidella sp. 2	R	605–1,300	Hard	LA, LH			
Isidid sp. A	R	1,448–1,475	Hard	CS			F. M. Bayer, pers. comm., 1996
Keratoisis sp. 1	R	550–740	Hard	LH			
Keratoisis sp. 2	O	458–2,012	Basalt	CS, LH, LV			
Keratoisis sp. 3	O	367–1,910	Basalt, talus	CS, WO			
Keratoisis sp. 4	C	710–1,870	Basalt, talus	CS, LH, LV			
Lepidisis olapa Musik	A	318–2,040	Hard	CS, HA	215–665	Hawaiian Is.	Musik 1978
Lepidisis sp. 1	R	1,400–1,700	Basalt	LV			
Lepidisis sp. 2	R	396–505	Basalt	LH			
Order Pennatulacea							
Sea pen yellow	R	1,870–1,900	Fine sediment	LH			Kolliker 1880
Family Funiculinidae							
Funiculina sp.	R	254–1,940	Fine sediment	CS, HI			
Family Pennatulidae							
Pennatula flava Nutting	O	257–367	Fine sediment	LH, M, WH	223–402	Hawaiian Is.	Nutting 1908
Pennatulid gray	R	248–575	Fine sediment	LH			
Pennatulid long	R	217–239	Fine sediment	WH			
Family Umbellulidae							
Umbellula sp.	R	680–1,900	Sand	LH, PB			
Family Kophobelemnidae							
Calibelemnon symmetricum Nutting	A	196–1,650	Hard	CS, HA, JA	153–1,274	Hawaiian Is.	Nutting 1908
Family Pteroeidae							
Pteroeides sp.	R	223–312	Fine sediment	LO, PB, WO			

Taxon		Depth	Substrate	Location	Reference
Family Virgulariidae					
Virgularia sp.	O	226–575	Fine sediment	HI, MO, WO	
Family Protoptilidae					
Protoptilum sp.	C	410–560	Fine sediment	CS	S. C. France, pers. comm.
Order Ceriantharia					
Cerianthid black	R	1,865–1,900	Sand	PB	
Cerianthid brown	O	76–367	Fine sediment	HI	
Cerianthid gray	O	168–367	Fine sediment	HI, JA	Hertwig 1888, Dunn 1982
Order Actiniaria					
Anemone brown	C	168–1,280	Hard	CS, LH, WH	
Anemone light brown	O	346–1,380	Hard	CS, HI, JA	
Anemone long	R	700–1,230	Basalt	CS, LH	
Anemone orange	O	1,770–1,800	Hard	MO	
Anemone rust	O	177–373	Hard	HI	
Anemone twin	O	650–700	Hard	LH	
Anemone white	O	199–1,800	Hard	CS, HA, JA	
Anemone white-tipped	C	281–500	Hard	CS, HI, JA	
Anemone wide base	R	915–1,025	Basalt	LH	
Anemone yellow	O	336–1,260	Hard	CS	
Family Actinostolidae					
Actinostolid tan	O	550–1,925	Hard	CS, HI	
Family Actinernidae					
Actinernus sp.	O	790–1,900	Basalt	CS, LH	
Family Actinoscyphiidae					
Actinoscyphia sp. 1	R	199–367	Hard	PB	
Actinoscyphia sp. 2	R	170–232	Gorgonian	LH, PB	
Actinoscyphia sp. 3	O	500–1,695	Gorgonian	CS, LH, LV	
Family Hormathiidae					
Hormathiid sp. 1	O	336–450	Gorgonian	LH, WO	
Hormathiid sp. 2	O	487–1,865	Cnidarian	CS, HI	
Hormathiid sp. 3	R	570–1,900	Basalt	LH, WH	
Hormathiid sp. 4	O	1,200–1,740	Cnidarian	LH	
Family Liponematidae					
Liponema brevicornis McMurrich	O	309–820	Talus	LH, WH	219–804 Pacific Dunn and Backus 1977
Family Nemanthidae					
Nemanthus sp.	R	122–1,755	Gorgonian	LH, M	
Order Corallimorpharia					
Corallimorphus sp.	O	95–1,675	Hard	CS, HI	Carlgren 1949
Order Zoanthinaria					
Zoanthid blue	O	352–415	Zoanthid	CS, LH, WH	

(continued on next page)

Table 1
A Taxonomic List of Animals Photographed from HURL Submersibles (continued)

TAXON	Rank* (observed)	Depth (m) (observed)	Substrate (on/above)	Location** (observed)	Depth (m) (published)	Location (published)	Selected References (depth, location, and/or photograph)
Zoanthid tan	R	500–1,910	Antipatharian	LH, LV			
Gerardia sp.	A	343–1,500	Gorgonian	CS, HI			
Parazoanthus sp. 1	O	343–460	Gorgonian	LH, LO, WH, WO			
Parazoanthus sp. 2	A	332–1,025	Gorgonian	LH, LO, WH, WO			
Order Scleractinia							
Scleractinian white-stalked	C	214–410	Limestone	JA, NWHI			
Family Oculinidae							
Madrepora sp.	R	665–1,010	Basalt	LH			
Family Acroporidae							
Acropora spp.	O	43–76	Limestone	JA			Maragos and Jokiel 1986
Montipora spp.	A	35–158	Hard	HA, JA			Maragos 1977, Maragos and Jokiel 1986
Family Agariciidae							
Leptoseris spp.	A	54–168	Limestone	HI, JA			Maragos 1977
Pavona sp.	R	57–59	Limestone	PB			
Family Pocilloporidae							
Madracis kauaiensis Vaughan	A	122–268	Hard	HI, JA	44–541	Hawaiian Is.	Vaughan 1907, Cairns 1984
Pocillopora spp.	O	37–73	Hard	HA, JA			Maragos 1977
Family Poritidae							
Porites spp.	A	15–187	Hard	HA, JA			Maragos 1977, Maragos and Jokiel 1986
Family Thamastreidae							
Psammocora sp.	R	86–122	Limestone	JA			Maragos 1977
Family Fungiidae							
Diaseris distorta (Michelin)	R	86–180	Coarse sediment	HA	17–475	Indo-Pacific	Cairns 1984
Family Caryophylliidae							
Caryophyllia rugosa Moseley	O	281–367	Hard	LH, PB, WH	71–439	Pacific	Cairns 1984
Trochocyathus sp.	R	305–335	Hard	LA			
Family Flabellidae							
Flabellum pavoninum Lesson	R	323–367	Fine sediment	M, LH	78–949	Indo-Pacific	Cairns 1984
Javania insignis Duncan	R	171–303	Limestone	HA	52–825	Indo-Pacific	Cairns 1984
Javania lamprotichum (Moseley)	O	251–440	Hard	CS, HA, JA	244–322	Central Pacific	Cairns 1984, Grigg et al. 1987
Family Dendrophylliidae							
Balanophyllia laysanensis Vaughan	R	336–339	Basalt	WH	237–270	Hawaiian Is.	Vaughan 1907

Species	Abundance	Depth (m)	Substrate	Location	Depth (m)	Region	References
Dendrophyllia oahuensis Vaughan	O	352–398	Limestone	JA, PB	357–485	Hawaiian Is.	Cairns 1984, Grigg et al. 1987
Dendrophyllia serpentina Vaughan	O	349–352	Limestone	WH	200–362	Hawaiian Is.	Cairns 1984
Dendrophyllid yellow	C	138–306	Limestone	HI			
Enallopsammia rostrata (Pourtalès)	C	505–1,200	Basalt	CS, HA	229–2,165	Circumglobal	Cairns 1984, Grigg et al. 1987
Order Antipatharia							
Antipathes dichotoma Pallas	O	76–159	Limestone	HA, JA	15–100	Indo-Pacific	Grigg and Eldredge 1975, Grigg 1977
Antipathes intermedia (Brook)	O	287–413	Hard	HA, JA	347–366	Hawaiian Is.	Grigg and Opresko 1977
Antipathes punctata (Roule)	R	330–367	Limestone	JA, LO, WH	300–445	Hawaiian Is.	Grigg and Opresko 1977
Antipathes sp. 1	C	315–400	Basalt	LO, WH, WO			
Antipathes sp. 2	R	420–480	Basalt	CS, WH			
Antipathes subpinnata Ellis and Solander	O	125–306	Limestone	JA, WO	455–460	Hawaiian Is.	Grigg and Opresko 1977
Antipathes ulex Ellis and Solander	O	157–364	Basalt	WH	30–330	Central Pacific	Grigg and Eldredge 1975, Grigg and Opresko 1977
Antipathes undulata van Pesch	R	355	Limestone	LO	110–490	Hawaiian Is.	Grigg and Opresko 1977
Bathypathes conferta (Brook)	C	306–1,745	Hard	CS, HI, JA	343–380	Hawaiian Is.	Brook 1889, Grigg and Opresko 1977
Bathypathes patula (Brook)	C	355–1,925	Hard	HI	458–4,830	Circumglobal	Opresko 1974, Pasternak 1977
Cirrhipathes anguinea Dana	C	40–153	Hard	HI, JA	15–46	Indo-Pacific	Grigg and Eldredge 1975, Grigg and Opresko 1977, Russo 1994
Cirrhipathes spiralis (Linnaeus)	A	107–450	Hard	HA, JA	124–454	Pacific	Grigg and Eldredge 1975, Grigg and Opresko 1977, Ohta 1983
Leiopathes glaberrima (Esper)	O	297–450	Hard	HA, JA	176–549	Circumglobal	Opresko 1974, Grigg and Opresko 1977
Parantipathes sp.	O	336–382	Hard	HI			
Schizopathes sp.	R	1,960–1,975	Fine sand	PB			
Stichopathes echinulosus Brook	O	800–1,690	Basalt	LA, PB	305–1,350	Pacific	Brook 1889, Grigg and Opresko 1977
Phylum Echinodermata							
Class Asteroidea							
Family Astropectinidae							
Astropectin pusillulus Fisher	R	232–257	Fine sediment	M, PB	256–683	Indo-Pacific	Fisher 1906, A. M. Clark 1989
Astropectinoides callistus Fisher	R	232–367	Fine sediment	M, PB, WH	97–426	Hawaiian Is.	Fisher 1906, A. M. Clark 1989
Tritonaster craspedotus Fisher	R	312–326	Fine sediment	LO, M	402–572	Hawaiian Is.	Fisher 1906, A. M. Clark 1989
Family Goniasteridae							
Anthenoides epixanthus (Fisher)	O	328–413	Fine sediment	LO, WH, WO	110–488	Pacific	Fisher 1906, A. M. Clark 1993

(continued on next page)

Table 1
A Taxonomic List of Animals Photographed from HURL Submersibles (continued)

TAXON	Rank* (observed)	Depth (m) (observed)	Substrate (on/above)	Location** (observed)	Depth (m) (published)	Location (published)	Selected References (depth, location, and/or photograph)
Calliaster pedicellaris Fisher	O	352–740	Cnidarian	LH	237–276	Hawaiian Is.	Fisher 1906, Grigg et al. 1987, A. M. Clark 1993
Calliderma spectabilis Fisher	O	138–274	Fine sediment	HA	143–387	Hawaiian Is.	Fisher 1906, A. M. Clark 1993
Ceramaster bowersi (Fisher)	R	244–367	Hard, sand	M	411–593	Hawaiian Is.	Fisher 1906, A. M. Clark 1993
Cryptopeltaster sp.	O	335–416	Limestone, sand	WO			
Mediaster ornatus Fisher	O	278–1,160	Hard	CS, LH, WH	523–1,360	Indo-Pacific	Fisher 1906, A. H. Clark 1949, A. M. Clark 1993
Plinthaster ceramoideus Fisher	O	294–413	Hard	HI	420–510	Hawaiian Is.	Fisher 1906, A. M. Clark 1993
Sphaeriodiscus ammophilus (Fisher)	O	245–400	Hard, sand	CS, HI	402–470	Hawaiian Is.	Fisher 1906, A. H. Clark 1949, A. M. Clark 1993
Family Asterodiscidae							
Asterodiscides tuberculosus (Fisher)	R	1,150–1,375	Basalt	LH	59–396	Hawaiian Is.	A. H. Clark 1949, A. M. Clark 1993
Family Ophidiasteridae							
Leiaster leachii Gray	R	49–168	Limestone	HI	10–30	Indo-Pacific	Fisher 1906, A. M. Clark 1993
Linckia diplax (Müller and Troschel)	O	46–138	Limestone	HA	18–78	Indo-Pacific	Sladen 1889, Fisher 1906, A. M. Clark 1993
Linckia multifora (Lamarck)	O	40–232	Hard	HA, JA	1–24	Indo-Pacific	Fisher 1906, A. H. Clark 1949, A. M. Clark and Rowe 1971
Ophidiaster rhabdotus Fisher	R	153–382	Hard	HA	426–438	Hawaiian Is.	Fisher 1906, A. M. Clark 1993
Tamaria scleroderma (Fisher)	R	339–358	Hard	M, LO, WO	181–193	Hawaiian Is.	Fisher 1906, A. M. Clark 1993
Tamaria tenella (Fisher)	R	365–367	Sand	M	237–276	Hawaiian Is.	Fisher 1906, A. M. Clark 1993
Tamaria triseriata (Fisher)	R	174–257	Limestone	JA, LH, PB	124–164	Hawaiian Is.	Fisher 1906, A. M. Clark 1993
Family Oreasteridae							
Culcita novaeguineae Müller and Troschel	R	168–220	Limestone	HA	3–18	Indo-Pacific	Sladen 1889, A. M. Clark and Rowe 1971, Russo 1994
Pentaceraster cumingi (Gray)	O	52–132	Coarse sediment	HA, JA	58–73	Pacific	Fisher 1906, Ely 1942, A. M. Clark 1993
Family Asterinidae							
Anseropoda insignis Fisher	R	260–275	Fine sediment	LH, M	223–332	Hawaiian Is.	Fisher 1906
Family Echinasteridae							
Henricia pauperrima Fisher	O	229–1,240	Limestone	HI, JA	362–1,464	Pacific	A. H. Clark 1949, Ohta 1983
Henricia robusta Fisher	R	281–382	Hard, sponge	JA, PB	349–441	Pacific	Fisher 1906
Family Pterasteridae							
Hymenaster pentagonalis Fisher	R	1,105–1,800	Hard	CS, MO	529–617	Hawaiian Is.	Fisher 1906

Taxon		Depth	Substrate	Codes	Depth	Region	Reference
Family Solasteridae							
Solaster sp.	R	336–1,540	Sand	CS, LH, WH			
Family Asteriidae							
Coronaster eclipes Fisher	O	245–805	Sand, talus	HA, JA	27–146	Hawaiian Is.	Fisher 1925
Schlerasterias euplecta (Fisher)	R	190–367	Limestone	LH, M, WO	89–362	Hawaiian Is.	Fisher 1906, 1925
Family Brisingidae							
Brisinga sp. 1	O	367–1,940	Hard	CS, LV, WH			
Brisinga sp. 2	R	750–1,975	Basalt	LH, PB			
Brisinga sp. 3	R	1,500–1,640	Limestone	LH, PB			
Brisingid white	R	1,130	Sand	LH			
Novodinia pacifica (Fisher)	R	605–1,475	Hard	WH	513–966	Hawaiian Is.	Fisher 1906
Family Zoroasteridae							
Zoroaster sp	R	1,620–1,680	Hard	LH			
Class Crinoidea							
Family Bathycrinidae							
Bathycrinus sp.	O	1,500–1,975	Basalt	HI, LV			
Family Hyocrinidae							
Diplocrinus sp.	O	960–1,400	Basalt	LH			
Family Porphyrocrinidae							
Porphyrocrinus sp.	O	1,432–1,960	Hard	LH			
Proisocrinus ruberrimus (Clark)	O	970–1,865	Hard	CS, LH, PB	929–1,700	Pacific	Macurda and Meyer 1976, Roux 1994
Family Hyocrinidae							
Ptilocrinus sp.	R	1,640–1,975	Basalt	LH, PB			
Family Antedontidae							
Antedon sp. (yellow)	O	850–1,740	Stalks	CS, LH			
Family Atelecrinidae							
Atelecrinus conifer Clark	O	665–1,960	Hard, talus	CS, LH, LV	1,009–1,479	Hawaiian Is.	A. H. Clark 1908
Family Charitometridae							
Charitometrid banded	R	550–1,700	Talus, stalks	CS, HA			
Family Comatulidae							
Comatulid brown	O	775–2,012	Hard	CS, LH			
Comatulid tan	O	410–2,040	Various	CS, LH			
Family Thalassometridae							
Cosmiometra crassicirra (Clark)	O	229–312	Limestone, stalks	HA	248–649	Hawaiian Is.	A. H. Clark 1908
Thalassometrid black	O	785–1,025	Stalks	CS, LH			
Thalassometrid yellow	O	1,325–1,865	Basalt	PB			
Class Ophiuroidea							
Ophiuroid tan	O	580–805	Fine sand	LA, MO			

(continued on next page)

Table 1
A Taxonomic List of Animals Photographed from HURL Submersibles (continued)

TAXON	Rank* (observed)	Depth (m) (observed)	Substrate (on/above)	Location** (observed)	Depth (m) (published)	Location (published)	Selected References (depth, location, and/or photograph)
Family Gorgonocephalidae							
Gorgonocephalid	O	394–1,535	Basalt, stalks	CS, LH, WO			
Family Asteroschematidae							
Astroschema caudatum (Lyman)	O	346–352	Gorgonian	WH	539–2,462	Pacific	A. H. Clark 1949; D. M. Devaney, pers. comm., 1984
Astroschema sp.	A	480–2,012	Gorgonian	CS, HI, LV			
Family Ophiomyxidae							
Ophiomyxa fisheri Clark	O	321–382	Fine sediment, coarse sediment	JA, LH, LO, WH, WO	497–523	Hawaiian Is.	A. H. Clark 1949; D. M. Devaney, pers. comm., 1984
Family Amphiuridae							
Histampica cythera (Clark)	O	229–260	Sand	LO, M, WO	90–322	Hawaiian Is.	A. H. Clark 1949; D. M. Devaney, pers. comm., 1984
Family Ophiothricidae							
Macrophiothrix lepidus (Clark)	O	205–245	Antipatharian	JA	45–298	Hawaiian Is.	A. H. Clark 1949; D. M. Devaney, pers. comm., 1984
Class Holothuroidea							
Family Elipidiidae							
Peniagone wyvillii Théel	R	1,295–1,925	Talus, sand	LV	2,435–4,413	Central Pacific	Théel 1882, Hansen 1975
Family Deimatidae							
Orphnurgus glaber (Walsh)	R	815–990	Sand	LH	245–1,210	Indo-Pacific	Hansen 1975
Family Laetmogonidae							
Pannychia moseleyi Théel	R	915–930	Fine sediment	CS	212–2,598	Pacific	Hansen 1975
Family Pelagothuriidae							
Enypniastes eximia Théel	R	915–1,940	Midwater	LH, LO	380–2,012	Pacific	Théel 1882, Ohta 1983
Family Psychoprotidae							
Benthodytes sp.	R	1,908–1,920	Fine sediment	PB			
Family Holothuriidae							
Actinopyga mauritiana (Quoy and Gaimard)	R	57–122	Hard, sand	JA, PB	Shallow	Indo-Pacific	Fisher 1907, A. M. Clark and Rowe 1971, Russo 1994
Actinopyga obesa Selenka	R	76–89	Coarse sediment	PB	Shallow	Hawaiian Is.	Fisher 1907
Holothuria arenicola Semper	C	54–116	Coarse sediment	LH, PB	1–9	Circumglobal	Fisher 1907, A. M. Clark and Rowe 1971, Maluf 1988
Holothuria atra Jaeger	O	15–113	Sand	HA, JA	Shallow	Indo-Pacific	A. M. Clark and Rowe 1971, Rowe and Doty 1977, Russo 1994
Holothuria spp.	O	50–60	Limestone	PB			

Taxon		Depth (m)	Substrate	Location code	Depth range	Region	Reference
Family Synallactidae							
Bathyplotes patagiatus Fisher	R	1,025–1,530	Talus, sand	LV	301–910	Hawaiian Is.	Fisher 1907
Mesothuria carnosa Fisher	R	1,550–1,800	Fine sediment	LH	356–936	Hawaiian Is.	Fisher 1907
Mesothuria murrayi Théel	R	400–410	Sand	LH	400–2,514	Indo-Pacific	Théel 1882, Fisher 1907
Mesothuria parva Fisher	R	910	Sand, basalt	CS	274–784	Pacific	Théel 1882, Fisher 1907
Paelopatides retifer Fisher	O	398–1,975	Fine sediment	LH, JA	405–1,479	Hawaiian Is.	Fisher 1907
Pseudostichopus propinquus Fisher	R	975–1,000	Fine sediment	LH	518–519	Hawaiian Is.	Fisher 1907
Synallactes sp.	R	1,150–1,170	Basalt	LV			
Family Stichopodidae							
Stichopus horrens Selenka	R	54–153	Sand, talus	LO, PB, WO	1–20	Indo-Pacific	Théel 1886, Rowe and Doty 1977, Maluf 1988
Class Echinoidea							
Family Cidaridae							
Acanthocidaris hastigera Agassiz and Clark	O	160–263	Limestone	HI	57–355	Central Pacific	Agassiz and Clark 1908, Mortensen 1928, Tabachnick 1981
Actinocidaris thomasii (Agassiz and Clark)	C	110–291	Limestone	HA, JA	128–400	Hawaiian Is.	Agassiz and Clark 1908, Mortensen 1928
Chondrocidaris gigantea Agassiz.	C	38–177	Hard	HI, JA	25–393	Indo-Pacific	Agassiz and Clark 1908, Mortensen 1928, Devaney 1987
Histocidaris variabilis (Agassiz and Clark)	R	138–550	Hard	LH, LO, WH, WO	320–555	Hawaiian Is.	Agassiz and Clark 1908, Mortensen 1928
Prionocidaris hawaiiensis (Agassiz and Clark)	R	92–214	Coarse sediment	LH, PB	32–510	Hawaiian Is.	Agassiz and Clark 1908, Mortensen 1928
Stereocidaris hawaiiensis Mortensen	C	257–490	Hard	HA	285–555	Hawaiian Is.	Agassiz and Clark 1908, Mortensen 1928
Stylocidaris calacantha (Agassiz and Clark)	O	171–303	Limestone	HI	225–355	Hawaiian Is.	Agassiz and Clark 1908, Mortensen 1928
Stylocidaris rufa Mortensen	C	98–367	Limestone, talus	JA, LH, LO, PB, WH, WO	128–515	Hawaiian Is.	Mortensen 1928
Family Aspidodiadematidae							
Aspidodiadema arcitum Mortensen	O	525–1,800	Hard, sand	CS, HI	300–1,450	Hawaiian Is.	Mortensen 1940
Aspidodiadema sp.	O	975–1,985	Hard, sand	CS, LH			
Family Diadematidae							
Astropyga radiata (Leske)	R	49–120	Coarse sediment	LH, PB	2–60	Indo-Pacific	Mortensen 1940, A. M. Clark and Rowe 1971
Centrostephanus asteriscus Agassiz and Clark	R	365–367	Hard	HA	60–315	Pacific	Agassiz and Clark 1908, Mortensen 1940
Chaetodiadema pallidum Agassiz and Clark	O	144–840	Fine sediment	HA	50–402	Hawaiian Is.	Agassiz and Clark 1908, Mortensen 1940

(continued on next page)

Table 1
A Taxonomic List of Animals Photographed from HURL Submersibles (continued)

TAXON	Rank* (observed)	Depth (m) (observed)	Substrate (on/above)	Location** (observed)	Depth (m) (published)	Location (published)	Selected References (depth, location, and/or photograph)[†]
Diadema savignyi Michelin	C	40–220	Hard	HA, JA	2–70	Pacific	Agassiz and Clark 1908, Mortensen 1940, A. M. Clark and Rowe 1971
Echinothrix diadema (Linnaeus)	R	61–122	Hard	LO, PB, WO	1–90	Indo-Pacific	Agassiz and Clark 1908, Mortensen 1940, A. M. Clark and Rowe 1971
Family Lissodiadematidae							
Lissodiadema purpureum (Agassiz and Clark)	R	205–263	Limestone, talus	JA, M, PB	40–82	Pacific	Agassiz and Clark 1908, Mortensen 1940
Family Toxopneustidae							
Tripneustes gratilla (Linnaeus)	R	45–63	Hard	HA	1–75	Indo-Pacific	Mortensen 1940, A. M. Clark and Rowe 1971
Family Pedinidae							
Caenopedina pulchella (Agassiz and Clark)	O	291–460	Hard	HA, JA	400–900	Pacific	Agassiz and Clark 1908, Mortensen 1940
Family Echinothuridae							
Echinothurid gray	R	1,090–1,380	Sand	CS, LV			
Echinothurid red	O	177–281	Hard	HI			
Echinothurid white	R	750–1,460	Sand	LH			
Phormosoma bursarium Agassiz	O	530–735	Sand	LA, MO	301–1,214	Pacific	Agassiz and Clark 1908
Sperosoma obscurum Agassiz and Clark	R	1,235–1,935	Sand	LH	301–1,245	Hawaiian Is.	Agassiz and Clark 1908, Mortensen 1940
Family Asterostomatidae							
Eurypatagus ovalis Mortensen	C	98–211	Limestone	HI, JA	190–192	Pacific	Mortensen 1950
Family Palaeopneustidae							
Phrissocystis multispina Agassiz and Clark	R	1,550–1,800	Sand	LH	Unknown	Hawaiian Is.	H. L. Clark 1917, Mortensen 1950, Eckelbarger et al. 1989
Family Laganidae							
Laganum fudsiyama Doderlein	O	297–370	Fine sediment	LH, LO, WO	51–643	Pacific	H. L. Clark 1914, Mortensen 1948, Ohta 1983
Phylum Arthropoda							
Order Decapoda							
Crab bumpy	R	246–263	Limestone	JA			
Crab white	O	945–1,925	Hard	LH, LV			
Shrimp red	A	367–1,950	Hard, sand	HA			
Shrimp striped	R	1,090–1,660	Hard, sand	CS, LH			

Taxon	R/O	Depth	Substrate	Location	Depth 2	Region	References
Family Aristeidae							
Aristaeus semidentatus (Bate)	R	1,030–1,870	Sand	LA, LH, PB	416–1,479	Indo-Pacific	Rathbun 1906, Burkenroad 1936
Family Bresiliidae							
Opaepele loihi Williams and Dobbs	O	975–1,060	Talus, vents	LV	775–1,060	Hawaiian Is.	Williams and Dobbs 1995
Family Ophlophoridae							
Acanthephyra eximia Smith	R	700–900	Hard	LH	1,000–2,118	Indo-Pacific	Rathbun 1906, Moffitt and Parrish 1992
Family Penaeidae							
Plesiopenaeus edwardsianus (Johnson)	O	1,652–1,920	Sand	LH, PB	275–1,850	Circumglobal	Burukovskii 1980, Takeda and Okutani 1983, Freitas 1985
Family Nematocarcinidae							
Nematocarcinus tenuirostris Bate	O	875–1,425	Hard	LH	500–3,000	Indo-Pacific	deMan 1920, Holthuis 1955
Family Pandalidae							
Heterocarpus ensifer Milne-Edwards	O	346–500	Hard, sand	HI	145–1,280	Circumglobal	Takeda and Okutani 1983, Chace 1985, Gooding et al. 1988, Moffitt and Parrish 1992
Heterocarpus laevigatus Bate	O	367–1,050	Hard, sand	LH, LO, WO	365–966	Indo-Pacific	Clarke 1972, Chace 1985, Gooding et al. 1988, Moffitt and Parrish 1992
Plesionika alcocki (Anderson)	O	298–850	Hard	HA	110–1,412	Indo-Pacific	Clarke 1972, Ohta 1983, Moffitt and Parrish 1992
Plesionika edwardsii (Brandt)	O	346–745	Hard	HA	150–880	Circumglobal	Kensley 1981, Takeda and Okutani 1983, Chace 1985
Plesionika ensis (Milne-Edwards)	R	367–900	Hard	LO	230–732	Circumglobal	Clarke 1972, Chace 1985, Moffitt and Parrish 1992
Plesionika martia (Milne-Edwards)	O	245–367	Limestone	LH, LO, WO	110–687	Circumglobal	Holthuis 1955, Clarke 1972, Ohta 1983, Chace 1985
Plesionika pacifica (Edmondson)	O	330–391	Hard	HI	110–220	Hawaiian Is.	Edmondson 1952, Clarke 1972
Plesionika sp.	O	202–1,335	Hard, talus	HI			
Family Hippolytidae							
Lysmata amboinensis (deMan)	R	61–177	Talus	HA	1–270	Pacific	Rathbun 1906, Debelius 1984
Family Stenopodidae							
Stenopus hispidus (Olivier)	R	92–110	Hard	PB	Shallow	Circumglobal	Bate 1888, Kensley 1981, Russo 1994
Stenopus pyrsonotus Goy and Devaney	O	83–229	Hard	HI, JA	23–68	Indo-Pacific	Goy and Devaney 1980, Russo 1994
Family Latreillidae							
Latreillia metanesa (Williams)	R	250–255	Sand	JA	44–278	Central Pacific	Williams 1982
Family Leucosiidae							
Randallia distincta Rathbun	O	266–373	Hard, sand	HI, JA	261–643	Hawaiian Is.	Rathbun 1906, Clarke 1972

(continued on next page)

Table 1
A Taxonomic List of Animals Photographed from HURL Submersibles (continued)

TAXON	Rank* (observed)	Depth (m) (observed)	Substrate (on/above)	Location** (observed)	Depth (m) (published)	Location (published)	Selected References (depth, location, and/or photograph)
Family Cancridae							
Cancer macrophthalmus (Rathbun)	R	269–343	Hard	LO	350–462	Hawaiian Is.	Rathbun 1906, Takeda 1977
Family Calappidae							
Mursia hawaiiensis Rathbun	R	289–360	Fine sediment	LO, WH	96–729	Hawaiian Is.	Rathbun 1906, Clarke 1972
Family Majidae							
Cyrtomaia smithi Rathbun	O	309–700	Limestone, sand	JA, LO, WH, WO	100–1,463	Hawaiian Is.	Rathbun 1906, Clarke 1972
Family Dynomenidae							
Dynomene devaneyi Takeda	O	336–373	Hard	HI, JA	350–365	Hawaiian Is.	Takeda 1977
Family Dromiidae							
Dromia dormia (Linnaeus)	R	61–156	Hard, sand	LH	1–50	Indo-Pacific	Kensley 1981
Family Lithodidae							
Lithodes longispinna Sakai	R	1,090–1,300	Hard	LH	550–1,250	Pacific	Hiramoto 1974, Takeda 1974, T. Sakai 1978
Lithodes nintokuae Sakai	R	336–785	Hard	CS	622–1,070	Pacific	Dawson and Yaldwyn 1985, K. Sakai 1987
Lithodes sp.	R	260–367	Hard	HA			
Neolithodes sp.	R	1,000–1,280	Sand	LH, MO			
Family Homolidae							
Homola orientalis Henderson	O	235–373	Hard	HA	111–200	Indo-Pacific	Kensley 1981, Ohta 1983
Paromala alcocki Stebbing	O	298–500	Sand, hard	HA	80–800	Indo-Pacific	Kensley 1981, Ohta 1983
Paromala japonica Paris	C	122–398	Hard	HA	165–375	Hawaiian Is.	Clarke 1972
Family Parthenopidae							
Parthenope stellata Rathbun	R	245–367	Hard	HI, JA	95–570	Hawaiian Is.	Rathbun 1906
Family Chirostylidae							
Eumunida smithii (Henderson)	C	281–965	Gorgonians	CS, HI	370–380	Pacific	Grigg et al. 1987
Eumunida sp.	O	346–1,640	Cnidarians	LH, PB, WH			
Galatheid long-armed	O	750–1,360	Sponges, cnidarians	CS, LH, LV			
Family Galatheidae							
Munida brucei Baba	O	352–1,925	Gorgonians	HI, JA, LV	119–1,370	Indo-Pacific	Baba 1974
Munida hawaiiensis Baba	O	180–800	Hard	HA	115–439	Hawaiian Is.	Baba 1981
Munida heteracantha Ortmann	A	104–349	Hard	HA	30–460	Pacific	Ohta 1983, Titgen 1987

Species		Depth (m)	Substrate	Location	Range	Region	Reference
Munida normani Henderson	O	251–420	Fine sediment	HA	365–460	Central Pacific	Titgen 1987
Munida sp. 3 (pink)	R	153–220	Fine sediment	M, PB, WH			
Munidopsis sp.	O	460–1,870	Basalt	LV			
Family Paguridae							
Pagurid in *Cerithium* shell	O	346–359	Hard, sand	LO			
Pagurid in *Dentalium* shell	O	217–306	Sand	WO			
Pagurid in other mollusk shell	C	15–1,730	Hard, sand	CS, HA, JA			
Family Parapaguridae							
Parapagurus dofleini Balss	O	300–505	Hard, sand	HI, JA	246–596	Pacific	Dunn et al. 1980, Ohta 1983
Family Diogenidae							
Trizopagurus hawaiiensis McLaughlin and Bailey-Brock	R	183–367	Hard, sand	HA	172–382	Hawaiian Is.	McLaughlin and Bailey-Brock 1975
Family Panuliridae							
Panulirus marginatus (Quoy and Gaimard)	O	46–240	Hard	HA, JA	18–142	Hawaiian Is.	Rathbun 1906
Family Scyllaridae							
Scyllarides squammosus (Milne-Edwards)	O	38–183	Hard	PB	3–47	Hawaiian Is.	Rathbun 1906
Family Protosquillidae							
Echinosquilla sp.	R	395–750	Hard	LH			
Family Odontodactylidae							
Odontodactylus hawaiiensis Manning	R	150–183	Sand	HI	109–276	Hawaiian Is.	Manning 1967
Order Thoracica							
Alcockianum alcockianum Annandale	R	1,600–1,640	Stalks	LH	1,098–1,800	Indo-Pacific	Rao and Newman 1972, Wilson et al. 1985
Barnacle balanoid	R	120–785	Stalks	HA			
Barnacle gooseneck	O	500–1,740	Stalks	HA			
Barnacle pink	R	161–162	Limestone	HA			
Phylum Mollusca							
Family Mastigoteuthidae							
Mastigoteuthis sp.	R	870–880	Midwater	WH			
Family Ommastrephidae							
Nototodarus hawaiiensis (Berry)	R	349–530	Sand	LO, WH, WO	400–540	Central Pacific	Roper et al. 1984, R. E. Young 1995
Family Stauroteuthidae							
Grimpoteuthis sp.	R	1,500–1,576	Midwater, hard	CS			
Family Cirroteuthidae							
Cirroteuthis sp.	O	1,335–1,652	Midwater, hard	LV, WO			

(continued on next page)

Table 1
A Taxonomic List of Animals Photographed from HURL Submersibles (continued)

TAXON	Rank* (observed)	Depth (m) (observed)	Substrate (on/above)	Location** (observed)	Depth (m) (published)	Location (published)	Selected References (depth, location, and/or photograph)
Family Tremoctopodidae							
Tremoctopus sp.	R	500–510	Midwater, sand	LA	Midwater	Circumglobal	Nesis 1982, R. E. Young 1995
Family Alloposidae							
Haliphron atlanticus Steenstrup	R	500–520	Midwater, sand	LA	Midwater	Circumglobal	Nesis 1982, R. E. Young 1995
Family Octopodidae							
Berrya hoylei (Berry)	O	321–398	Hard	HI, JA	470–841	Hawaiian Is.	Berry 1914
Octopus cyanea Gray	R	109–220	Sand, hard	LO	1–50	Indo-Pacific	Kay 1979, Roper et al. 1984
Octopus ornatus Gould	R	229–324	Hard	JA, LO, WO	1–50	Hawaiian Is.	Kay 1979
Octopus sp.	O	162–404	Hard	LH, LO, WO			
Octopus sp. B	O	112–410	Hard	WH			
Scaeurgus patagiatus Berry	R	343–367	Hard	WH	232–326	Hawaiian Is.	Berry 1914
Family Dentaliidae							
Dentalium sp.	R	300–306	Sand	WO			
Family Pinnidae							
Pinna muricata Linnaeus	O	52–92	Hard, sand	LH, PB	25–250	Indo-Pacific	Kay 1979
Family Cassididae							
Cassis cornuta (Linnaeus)	R	49–92	Coarse sediment	PB	3–65	Indo-Pacific	Kay 1979
Family Cerithiidae							
Cerithium matukense Watson	R	352–372	Sand	LO	40–567	Pacific	Kay 1979
Family Xenophoridae							
Xenophora peroniana (Iredale)	R	268–321	Sand	LH, LO, WO	20–400	Pacific	Rehder and Ladd 1973, Kay 1979
Family Pleurobranchidae							
Pleurobranchus sp.	O	232–800	Hard, sand	HA			
Order Cephalaspida							
Cephalaspid	R	483–487	Sand	WO			
Family Tritoniidae							
Tritonia sp.	O	352–398	Hard, sponge	LH, WO			
Phylum Chordata							
Class Ascidiacea							
Family Perophoridae							
Perophora sp.	C	223–410	Hard	HA			Goodbody 1982
Class Chondrichthys							

Taxon	Abundance	Depth	Substrate	Location	Depth observed	Region	References
Family Hexanchidae							
Hexanchus griseus (Bonnaterre)	R	500–1,400	Midwater	WH, WO	183–1,524	Circumglobal	Compagno 1984, Humphreys et al. 1984, Borets 1986
Family Alopiidae							
Alopias vulpinus (Bonnaterre)	R	320	Midwater	JA	1–366	Circumglobal	Compagno 1984, Randall et al. 1985
Family Pseudotriakidae							
Pseudotriakis microdon Capello	R	500	Sand	LH	200–1,500	Circumglobal	Tinker 1978, Compagno 1984
Family Carcharhinidae							
Carcharhinus amblyrhynchos (Bleeker)	O	10–275	Midwater	JA	10–274	Indo-Pacific	Johnson 1984, Randall et al. 1985, Myers 1989, Randall 1996
Family Squalidae							
Centrophorus granulosus (Bloch and Schneider)	R	500	Sand	LO	100–1,200	Circumglobal	Compagno 1984, Masuda et al. 1984
Squalus mitsukurii Jordan and Snyder	R	336–416	Hard	LH, LO, WO	160–545	Pacific	Clarke 1972, Struhsaker 1973, Novikov et al. 1980, Masuda et al. 1984, Parin 1991
Family Echinorhinidae							
Echinorhinus cookei Pietschmann	R	360–420	Sand	CS, LH	11–424	Pacific	Compagno 1984, Humphreys et al. 1984, Golovan and Pakhorukov 1987, Parin 1991
Family Torpedinidae							
Torpedo sp.	R	332	Fine sand	WH	265–476	Hawaiian Is.	Struhsaker 1973
Family Dasyatidae							
Dasyatis brevis (Garman)	R	52–150	Sand	HA	10–148	Pacific	Tinker 1978, Nishida and Nakaya 1990
Dasyatis latus (Garman)	R	40–75	Sand	LO	40–214	Pacific	Struhsaker 1973, Nishida and Nakaya 1990
Family Hexatrygonidae							
Hexatrygon longirostra Chu and Meng	R	750–950	Sand, talus	LH	700–710	Pacific	Okamura and Kitajima 1984
Family Plesiobatididae							
Plesiobatis daviesi (Wallace)	O	287–780	Sand	LO, WH, WO	44–412	Indo-Pacific	Clarke 1972, Struhsaker 1973, Masuda et al. 1984, Nishida 1990
Family Chimaeridae							
Hydrolagus purpurescens (Gilbert)	R	1,150–1,580	Hard	CS, LV, PB	1,750–1,951	Central Pacific	Gilbert 1905, Novikov et al. 1980, Masuda et al. 1984
Family Rhinochimaeridae							
Rhinochimaera pacifica (Mitsukurii)	R	1,000–1,136	Sand	LA, LO	980–1,100	Pacific	Masuda et al. 1984

(continued on next page)

Table 1
A Taxonomic List of Animals Photographed from HURL Submersibles (continued)

TAXON	Rank* (observed)	Depth (m) (observed)	Substrate (on/above)	Location** (observed)	Depth (m) (published)	Location (published)	Selected References (depth, location, and/or photograph)
Class Osteichthyes							
Family Halosauridae							
Aldrovandia phalacra (Vaillant)	O	955–1,680	Sand	LH, PB	530–1,635	Circumglobal	McDowell 1973, Struhsaker 1973, Kaufmann et al. 1989, Chave and Jones 1991
Family Notocanthidae							
Notocanthus sp.	R	740–940	Sand	LH	768–796	Hawaiian Is.	B. C. Mundy, pers. comm., 1994
Family Muraenidae							
Gymnothorax berndti Snyder	O	128–303	Hard, holes	HA, JA	110–225	Indo-Pacific	Struhsaker 1973, Tinker 1978, Myers 1989
Gymnothorax nudivomer (Playfair and Günther)	O	61–214	Hard, holes	JA, LH, PB	30–165	Indo-Pacific	Clarke 1972, Randall et al. 1985, Myers 1989
Gymnothorax nuttingi Snyder	O	110–291	Hard, holes	HA, JA	165–300	Johnston Atoll, Hawaiian Is.	Clarke 1972, Randall et al. 1985
Family Synaphobranchidae							
Ilyophine eel	O	343–396	Hard, *Gerardia*	HI			
Synaphobranchus brevidorsalis Günther	O	1,175–1,530	Sand	CS, LH, LV	575–1,535	Circumglobal	Robins and Robins 1989
Synaphobranchus kaupi Johnson	R	810–1,725	Fine sand	CS, LH	131–2,196	Circumglobal	Robins and Robins 1989
Synaphobranchus sp.	O	1,175–1,400	Sand, hard	HA	550–1,975		
Family Ophichthidae							
Myrichthys maculosus (Cuvier)	R	69–262	Various	JA	1–262	Indo-Pacific	Randall et al. 1985, 1990; Myers 1989
Ophichthus kunaloa McCosker	O	350–382	Fine sand	LO, WO	220–250	Hawaiian Is.	McCosker 1979
Family Congridae							
Ariosoma marginatum (Vaillant and Sauvage)	O	214–490	Fine sand	CS, JA, WH	101–236	Hawaiian Is.	Struhsaker 1973, Humphreys et al. 1984
Bathyuroconger vicinus (Vaillant)	R	750–900	Fine sand	LH	229–1,318	Circumglobal	D. G. Smith 1989
Conger oligoporus Kanazawa	O	217–470	Hard, holes	JA, PB	110–365	Johnston Atoll, Hawaiian Is.	Clarke 1972, Struhsaker 1973; Randall et al. 1985
Family Nettastomatidae							
Nettastoma parviceps Günther	R	775–945	Sand	CS, LH	420–1,140	Indo-Pacific	Tinker 1978, Ohta 1983, Okamura and Kitajima 1984, Smith and Heemstra 1986

Family / Species		Depth (m)	Substrate	Islands	Depth range	Region	References
Family Aulopodidae							
Hime japonica (Günther)	R	343–510	Fine sand	CS, LH, WH	201–238	Pacific	Struhsaker 1973, Borets 1986, Parin and Kotlyar 1989
Family Chloropthalmidae							
Chlorophthalmus proridens Gilbert and Cramer	O	218–510	Fine sand	LA, LO, M, WH, WO	185–350	Central Pacific	Gilbert and Cramer 1892, Novikov et al. 1980
Family Ipnopidae							
Bathypterois atricolor Alcock	R	1,800–1,920	Fine sand	LH, PB	573–5,150	Indo-Pacific	Sulak 1977, Masuda et al. 1984, Jones and Sulak 1990
Bathypterois grallator (Goode and Bean)	R	1,140–1,460	Fine sand	LH	878–4,720	Circumglobal	Sulak 1977, Jones and Sulak 1990
Family Synodontidae							
Synodus doaki Russell and Cressey	R	52–260	Hard	WA	90–200	Pacific	Waples and Randall 1988
Synodus sp. 3	R	350–352	Sand	WO			
Family Myctophidae							
Benthosema fibulatum (Gilbert and Cramer)	O	330–375	Over sand	LA, WH	75–856	Indo-Pacific	Gilbert and Cramer 1892, Struhsaker 1973, Novikov et al. 1980, Reid et al. 1991
Diaphus chysorhynchus Gilbert and Cramer	O	340–390	Over sand	LA, WH	75–960	Pacific	Gilbert and Cramer 1892, Struhsaker 1973, Novikov et al. 1980, Reid et al. 1991
Family Bathygadidae							
Gadomus melanopterus Gilbert	R	840–1,350	Sand, talus	CS	799–1,416	Pacific	Howes and Crimmen 1990, Sazonov and Iwamoto 1992
Family Macrouridae							
Caelorinchus spilonotus Sazonov and Iwamoto	R	349–480	Sand	CS, M, WH	360–600	Pacific	Sazonov and Iwamoto 1992
Caelorinchus sp. 1	R	680–755	Fine sand	LH			
Caelorinchus sp. 2	R	500	Sand, talus	LA			
Coryphaenoides longicirrhus (Gilbert)	R	1,450–1,985	Talus	LA, LV	1,650–2,365	Central Pacific	Tinker 1978, Wilson et al. 1985
Hymenocephalus sp. 1	R	700–840	Fine sand	LH			
Nezumia propinqua (Gilbert and Cramer)	R	350–385	Fine sand	WH	219–870	Indo-Pacific	Okamura et al. 1982, Sazonov and Iwamoto 1992
Ventrifossa sp. 1	R	1,640–1,700	Fine sand	LH			
Family Moridae							
Antimora microlepis Bean	R	1,880–1,885	Hard	LV	1,602–2,403	Pacific	Gilbert 1905, Small 1981
Gadella molokaiensis Paulin	R	339–352	Hard, holes	WH	181–400	Hawaiian Is.	Struhsaker 1973, Paulin 1989
Laemonema rhodochir Gilbert	O	313–404	Hard, holes	LH, LO, WO	280–600	Pacific	Parin 1984

(continued on next page)

Table 1
A Taxonomic List of Animals Photographed from HURL Submersibles (continued)

TAXON	Rank* (observed)	Depth (m) (observed)	Substrate (on/above)	Location** (observed)	Depth (m) (published)	Location (published)	Selected References (depth, location, and/or photograph)
Physiculus grinnelli Jordan and Jordan	R	150–242	Hard, holes	LH, PB, WH	120–320	Johnston Atoll, Hawaiian Is.	Struhsaker 1973, Humphreys et al. 1984, Randall et al. 1985, Paulin 1989
Family Ophidiidae							
Brotula multibarbata Temminck and Schlegel	R	95–199	Hard, holes	HA	2–220	Pacific	Gosline 1953, Myers 1989
Pycnocraspedum armatum Gosline	R	750–1,345	Sand	CS, LH	201–335	Hawaiian Is.	Struhsaker 1973
Family Carapidae							
Pyramodon ventralis Smith and Radcliffe	R	367	Hard	LO	100–450	Indo-Pacific	Masuda et al. 1984
Family Lophiidae							
Lophiodes miacanthus (Gilbert)	O	294–700	Various	LO, WH, WO	417–571	Central Pacific	Struhsaker 1973, Novikov et al. 1980, Caruso 1981
Sladenia remiger Smith and Radcliffe	O	780–1,540	Hard	CS, LH, LV	1,200–1,294	Indo-Pacific	Caruso and Bullis 1976, Caruso 1985
Family Chaunacidae							
Chaunax fimbriatus Hilgendorf	R	1,247–1,985	Hard	CS, LV	1,300	Pacific	Humphreys et al. 1984, Masuda et al. 1984, Caruso 1989
Chaunax umbrinus Gilbert	O	333–385	Hard	HA	183–400	Hawaiian Is.	Clarke 1972, Struhsaker 1973, Novikov et al. 1980
Family Ogcocephalidae							
Malthopsis jordani Gilbert	R	234–520	Sand	LH, WO	210–430	Central Pacific	Struhsaker 1973, Okamura and Kitajima 1984
Family Ateleopodidae							
Ijimaia plicatellus (Gilbert)	O	336–367	Fine sand	LO, WH, WO	265–500	Hawaiian Is.	Struhsaker 1973, Grigg et al. 1987
Family Trachichthyidae							
Paratrachichthys prosthemius Jordan and Fowler	O	193–195	Hard, holes	PB	90–110	Pacific	Struhsaker 1973; Gon 1983, 1987
Paratrachichthys sp.	O	242	Hard, holes	NWHI			
Family Berycidae							
Beryx decadactylus Cuvier	O	346–805	Hard	CS, LO, WH, WO	200–760	Circumglobal	Busakhin 1982, Shcherbachev et al. 1985, Borets 1986, Grigg et al. 1987
Family Holocentridae							
Myripristis berndti Jordan and Evermann	O	46–159	Hard, holes	LH, LO, WH, WO	15–50	Indo-Pacific	Strasburg et al. 1968; Greenfield 1974; Randall 1985, 1996; Myers 1989

Species		Depth 1	Substrate	Locations	Depth 2	Region	References
Myripristis chryseres Jordan and Evermann	A	46–211	Hard, holes	HA, JA	12–240	Indo-Pacific	Randall et al. 1985, 1990; Ralston et al. 1986; Myers 1989; Randall 1996
Neoniphon aurolineatus (Lienard)	C	49–188	Hard, holes	JA, LH, WH	30–160	Indo-Pacific	Randall et al. 1985, 1990; Ralston et al. 1986; Myers 1989
Pristilepis oligolepis (Whitley)	O	102–281	Hard, holes	HA	14–345	Pacific	Randall 1981a; Randall et al. 1982, 1985
Sargocentron microstoma (Günther)	R	165–183	Hard, holes	JA	1–183	Indo-Pacific	Myers 1989, Randall et al. 1990, Randall 1996
Sargocentron spiniferum (Forskål)	R	89–122	Hard, holes	JA, LH	1–122	Indo-Pacific	Strasburg et al. 1968, Myers 1989, Randall et al. 1990
Sargocentron xantherythrum (Jordan and Evermann)	O	61–217	Hard, holes	JA, LH, PB, WH	1–101	Johnston Atoll, Hawaiian Is.	Struhsaker 1973, Randall 1985, Randall et al. 1985
Family Polymixiidae							
Polymixia berndti Gilbert	O	352–585	Sand	CS, HI	99–500	Indo-Pacific	Struhsaker 1973, Uchida and Tagami 1984, M. M. Smith and Heemstra 1986, Kotlyar 1993
Family Zeidae							
Stethopristes eos Gilbert	R	343–570	Fine sand	HI	380–622	Pacific	Struhsaker 1973, Humphreys et al. 1984, Parin 1991
Zenopsis nebulosus (Schlegel)	R	349–367	Hard	LO, WO	216–670	Pacific	Struhsaker 1973, Heemstra 1980, Okamura et al. 1982, Golovan and Pakhorukov 1987
Family Grammicolepidae							
Grammicolepis brachiusculus Poey	R	339	Hard, holes	WH	400–800	Circumglobal	Borets 1986, M. M. Smith and Heemstra 1986
Family Caproidae							
Antigonia (*A. eos*, *A. capros*)	C	199–367	Various	CS, HA, JA WH	330–644	Pacific	Parin and Borodulina 1986
Cyttomimus stelgis (Gilbert)	R	336	Fine sand				Gilbert 1905, Parin 1991
Family Aulostomidae							
Aulostomus chinensis (Linnaeus)	O	61–122	Various	JA, LH, PB	1–124	Indo-Pacific	Strasburg et al. 1968, Struhsaker 1973, Myers 1989
Family Syngnathidae							
Dunckerocampus baldwini Herald and Randall	R	125–128	Talus	WH	6–48	Hawaiian Is.	Herald and Randall 1972, Randall 1996
Family Macroramphosidae							
Macroramphosid	R	245–366	Hard, sand	WO			
Family Scorpaenidae							
Ectreposebastes imus Garman	R	570–775	Fine sand	LA	150–850	Circumglobal	Struhsaker 1973, Eschmeyer and Randall 1975

(continued on next page)

Table 1
A Taxonomic List of Animals Photographed from HURL Submersibles (continued)

TAXON	Rank* (observed)	Depth (m) (observed)	Substrate (on/above)	Location** (observed)	Depth (m) (published)	Location (published)	Selected References (depth, location, and/or photograph)
Neomerinthe rufescens (Gilbert)	O	153–385	Hard, sand	HI, JA	75–302	Hawaiian Is.	Struhsaker 1973, Eschmeyer and Randall 1975
Pontinus macrocephalus (Sauvage)	C	177–367	Hard	HA, JA	120–366	Johnston Atoll, Hawaiian Is.	Clarke 1972, Struhsaker 1973, Eschmeyer and Randall 1975, Randall et al. 1985
Pterois sphex Jordan and Evermann	R	83–95	Hard	PB	10–124	Hawaiian Is.	Struhsaker 1973, Randall 1985
Scorpaena (*S. colorata*, *S. pele*)	O	122–272	Hard	HA, JA	90–272	Hawaiian Is.	Eschmeyer and Randall 1975, Randall et al. 1985, Ralston et al. 1986
Scorpaenopsis altirostris Gilbert	R	159–190	Hard	LH	79–134	Hawaiian Is. Johnston Atoll	Eschmeyer and Randall 1975
Setarches guentheri Johnson	O	349–780	Hard	CS, HI	185–686	Circumglobal	Struhsaker 1973, Kanayama 1981, Ohta 1983, Parin 1991
Family Triglidae							
Satyrichthys engyceros (Günther)	O	193–394	Hard, sand	HA, JA	183–503	Central Pacific	Struhsaker 1973, Masuda et al. 1984, Randall et al. 1985, Borets 1986
Satyrichthys hians (Gilbert and Cramer)	R	550	Sand	LO	275–610	Pacific	Tinker 1978, Masuda et al. 1984, Okamura 1985
Family Bembridae							
Bembradium roseum Gilbert	O	309–373	Sand	LH, LO, M, WO	210–650	Indo-Pacific	Struhsaker 1973, Humphreys et al. 1984, Okamura 1985, Parin 1991
Family Acropomatidae							
Synagrops argyrea (Gilbert and Cramer)	O	346–630	Hard	LA, LH, WH	75–522	Hawaiian Is.	Fraser 1972, Struhsaker 1973
Family Serranidae							
Caprodon schlegeli (Günther)	R	197–251	Hard, holes	LH, NWHI	260–302	Pacific	Struhsaker 1973, Masuda et al. 1984, Uchida and Tagami 1984, Okamura 1985
Epinephelus quernus Seale	C	89–315	Various	HA, JA	18–350	Johnston Atoll, Hawaiian Is.	Uchida and Tagami 1984, Randall et al. 1985, Ralston et al. 1986, Randall and Heemstra 1991
Holanthias elizabethae (Fowler)	C	107–291	Hard, holes	HI, JA	155–262	Johnston Atoll, Hawaiian Is.	Strasburg et al. 1968, Madden 1973, Randall et al. 1985
Holanthias fuscipinnis (Jenkins)	A	89–260	Hard, holes	HA, JA	55–213	Johnston Atoll, Hawaiian Is.	Strasburg et al. 1968, Randall et al. 1985, Pyle 1991, Severns and Fiene-Severns 1993, Randall 1996

Species		Depth	Substrate	Codes	Range	Region	References
Liopropoma aurora (Jordan and Evermann)	R	162–183	Hard, holes	LH, LO, WO	49–183	Hawaiian Is.	Randall and Taylor 1988, Randall 1996
Luzonichthys earlei Randall	O	76–205	Hard, holes	LH, PB, WH	40–200	Indo-Pacific	Randall 1981b, 1996; Randall et al. 1985; Randall and McCosker 1992
Plectranthias helenae Randall	R	122–263	Hard	HI, JA	199–220	Johnston Atoll, Hawaiian Is.	Randall 1980b, Randall et al. 1985
Plectranthias kelloggi (Jordan and Evermann)	O	245–370	Hard, holes	HA	221–310	Pacific	Clarke 1972, Struhsaker 1973, Uchida and Tagami 1984
Pseudanthias bicolor (Randall)	R	50–104	Hard, holes	HA	5–68	Pacific	Uchida and Tagami 1984, Myers 1989, Randall et al. 1990
Pseudanthias fucinus (Randall and Ralston)	C	122–266	Hard, holes	HA, JA	135–280	Johnston Atoll, Hawaiian Is.	Randall and Ralston 1984, Randall et al. 1985, Ralston 1986
Pseudanthias hawaiiensis Randall	A	40–199	Hard, holes	HA	26–120	Pacific	Randall 1979, 1996; Hobson and Chave 1990
Pseudanthias thompsoni (Fowler)	C	49–188	Hard, holes	HA	5–150	Hawaiian Is.	Randall 1979, 1996; Myers 1989
Family Symphysanodontidae							
Symphysanodon maunaloae Anderson	C	131–398	Various	HA, JA	230–366	Pacific	Struhsaker 1973, Masuda et al. 1984, Uchida and Tagami 1984, Parin 1991
Symphysanodon typus Bleeker	A	80–245	Hard	HA	119–236	Pacific	Anderson 1970
Family Callanthiidae							
Callanthias sp.	O	171–360	Hard, holes	HA, JA	240–330	Johnston Atoll	Ralston et al. 1986
Grammatonotus laysanus Gilbert	O	175–367	Hard, holes	JA	240–354	Pacific	Humphreys et al. 1984, Randall et al. 1985, Ralston et al. 1986, Parin 1991
Family Priacanthidae							
Cookeolus japonicus (Cuvier)	A	131–306	Hard, holes	HA	60–400	Circumglobal	Struhsaker 1973, Humphreys et al. 1984, Okamura 1985, Borets 1986, Starnes 1988
Priacanthus alalaua Jordan and Evermann	O	55–223	Hard, holes	JA, PB	8–250	Pacific	Starnes 1988
Priacanthus meeki Jenkins	C	61–272	Various	HA	3–238	Hawaiian Is.	Okamoto and Kanenaka 1984, Fielding and Robinson 1987, Starnes 1988, Randall 1996
Family Apogonidae							
Apogon kallopterus Bleeker	O	49–158	Sand, hard	PB	1–45	Indo-Pacific	Fraser 1972, Chave 1978, Hobson 1984, Myers 1989, Randall et al. 1990, Randall 1996
Apogon maculiferus Garrett	C	61–153	Sand, hard	LH, PB	3–128	Hawaiian Is.	Struhsaker 1973; Randall 1985, 1996

(continued on next page)

Table 1
A Taxonomic List of Animals Photographed from HURL Submersibles (continued)

TAXON	Rank* (observed)	Depth (m) (observed)	Substrate (on/above)	Location** (observed)	Depth (m) (published)	Location (published)	Selected References (depth, location, and/or photograph)
Family Epigonidae							
Epigonus atherinoides (Gilbert)	O	500–540	Sand, hard	LA, LH	500–704	Pacific	Struhsaker 1973, Mochizuki and Shirakihara 1983, Humphreys et al. 1984, Parin 1991
Epigonus fragilis (Jordan and Jordan)	O	312–367	Hard	JA	120–366	Hawaiian Is.	Mayer 1974
Epigonus glossodontus Gon	O	366–520	Hard	LA, LH, WO	366	Hawaiian Is.	Gon 1985
Family Malacanthidae							
Malacanthus brevirostris Guichenot	R	30–61	Sand, hard	PB	14–45	Indo-Pacific	Myers 1989
Family Emmelichthyidae							
Erythrocles scintillans (Jordan and Thompson)	O	221–606	Various	JA, LH, LO, WH, WO	120–320	Pacific	Clarke 1972, Heemstra and Randall 1977, Ralston et al. 1986
Family Bramidae							
Eumegistus illustris Jordan and Jordan	R	306–520	Sand	LH	270–620	Indo-Pacific	Mead 1972, Okamura et al. 1982, Masuda et al. 1984, Ralston et al. 1986
Family Lutjanidae							
Aphareus furca (Lacépède)	R	46–92	Midwater	HA, JA	1–122	Indo-Pacific	Randall et al. 1985, Ralston et al. 1986, Myers 1989, Randall 1996
Aphareus rutilans Cuvier	O	61–199	Midwater	HI, JA	100–250	Indo-Pacific	Allen 1985, Randall et al. 1985, Ralston et al. 1986, M. M. Smith and Heemstra 1986
Aprion virescens Valenciennes	O	58–120	Various	LH, PB	46–100	Indo-Pacific	Strasburg et al. 1968, Okamoto and Kanenaka 1984, Randall et al. 1990, Randall 1996
Etelis carbunculus Cuvier	O	89–398	Various	HA, JA	90–366	Indo-Pacific	Struhsaker 1973, Uchida and Tagami 1984, Okamura 1985, Randall et al. 1985
Etelis coruscans Valenciennes	O	168–396	Midwater	HA, JA	100–357	Indo-Pacific	Anderson 1981, Masuda et al. 1984, Allen 1985, M. M. Smith and Heemstra 1986, Roux 1994
Lutjanus fulvus (Schneider)	R	46–128	Various	LH, WH	2–75	Indo-Pacific	Allen and Talbot 1985, Myers 1989, Randall et al. 1990
Lutjanus kasmira (Forskål)	A	28–174	Various	HI	2–275	Indo-Pacific	Allen and Talbot 1985; Randall 1985, 1996; Myers 1989, Randall et al. 1990

Species							
Pristipomoides auricilla (Jordan, Evermann and Tanaka)	R	160–352	Various	JA, PB	90–360	Indo-Pacific	Randall 1980a, Allen 1985, Randall et al. 1985
Pristipomoides filamentosus (Valenciennes)	C	52–343	Various	HA, JA	90–328	Indo-Pacific	Uchida and Tagami 1984, Allen 1985, Randall et al. 1985
Pristipomoides zonatus (Valenciennes)	O	171–352	Various	JA, PB, WH	70–293	Indo-Pacific	Uchida and Tagami 1984, Allen 1985, Randall et al. 1985
Randallichthys filamentosus (Fourmanoir)	R	193–380	Hard	LH, WH	152–293	Pacific	Randall 1981a, Masuda et al. 1984
Family Mullidae							
Mulloides vanicolensis (Valenciennes)	O	38–116	Sand	JA, LO, WO	15–113	Indo-Pacific	Myers 1989, Randall et al. 1990, Randall 1996
Parupeneus bifasciatus (Lacépède)	R	68–83	Sand, hard	LH, PB, WH	1–80	Indo-Pacific	Randall 1985, 1996; Myers 1989
Parupeneus chrysonemus (Jordan and Evermann)	O	46–183	Sand	HI, JA	20–124	Hawaiian Is.	Struhsaker 1973, Tinker 1978, Randall 1996
Parupeneus cyclostomus (Lacépède)	R	49–113	Sand	HI, JA	5–125	Indo-Pacific	Randall 1985, 1996; Ralston et al. 1986; Myers 1989; Randall et al. 1990
Parupeneus multifasciatus (Quoy and Gaimard)	C	15–161	Various	HI, JA	3–140	Pacific	Strasburg et al. 1968, Struhsaker 1973, Randall et al. 1990, Randall 1996
Parupeneus pleurostigma (Bennett)	O	61–120	Various	LH, PB	1–75	Indo-Pacific	Strasburg et al. 1968, Struhsaker 1973, Randall et al. 1990, Severns and Fiene-Severns 1993
Parupeneus porphyreus (Jenkins)	O	40–122	Various	LO, PB, WH	2–140	Hawaiian Is.	Struhsaker 1973, Okamoto and Kanenaka 1984, Randall 1996
Family Chaetodontidae							
Chaetodon auriga Forskål	R	37–61	Coral heads	JA	3–30	Indo-Pacific	Myers 1989, Randall et al. 1990, Randall 1996
Chaetodon fremblii Bennett	R	61–128	Various	LH, PB, WH	4–183	Hawaiian Is.	Struhsaker 1973, Randall 1996
Chaetodon kleinii Bloch	R	41–122	Various	PB	10–61	Indo-Pacific	Strasburg et al. 1968, Myers 1989, Randall et al. 1990
Chaetodon lunula (Lacépède)	R	31–70	Various	LH	1–158	Indo-Pacific	Strasburg et al. 1968, Myers 1989, Randall et al. 1990
Chaetodon miliaris Quoy and Gaimard	A	46–128	Various	HI, JA	1–250	Johnston Atoll, Hawaiian Is.	Strasburg et al. 1968, Okamoto and Kanenaka 1984, Randall et al. 1985, Randall 1996
Chaetodon modestus Schlegel	A	89–291	Various	HI, JA	10–255	Indo-Pacific	Allen 1979, Okamura 1985, Randall et al. 1985

(continued on next page)

Table 1
A Taxonomic List of Animals Photographed from HURL Submersibles (continued)

TAXON	Rank* (observed)	Depth (m) (observed)	Substrate (on/above)	Location** (observed)	Depth (m) (published)	Location (published)	Selected References (depth, location, and/or photograph)
Chaetodon multicinctus Garrett	R	43–114	Various	LH, LO, WO	5–62	Johnston Atoll, Hawaiian Is.	Allen 1979; Hobson 1984; Randall 1985, 1996; Randall et al. 1985
Chaetodon quadrimaculatus Gray	R	40–43	Various	JA	2–15	Central Pacific	Randall 1985, 1996; Randall et al. 1985
Chaetodon tinkeri Schultz	O	76–183	Various	HI, JA	40–160	Central Pacific	Randall 1985, 1996; Myers 1989
Forcipiger flavissimus Jordan and McGregor	C	15–128	Various	HI, JA	2–145	Indo-Pacific	Ralston et al. 1986, Myers 1989, Randall et al. 1990
Forcipiger longirostris (Broussonet)	O	43–208	Various	HI	5–60	Indo-Pacific	Strasburg et al. 1968, Myers 1989, Randall et al. 1990
Hemitaurichthys polylepis (Bleeker)	R	40–46	Various	WH	3–40	Pacific	Myers 1989, Randall et al. 1990, Randall 1996
Hemitaurichthys thompsoni Fowler	O	46–70	Various	JA	10–114	Central Pacific	Randall et al. 1985, Myers 1989, Randall 1996
Heniochus diphreutes Jordan	A	46–177	Various	HI, JA	3–215	Indo-Pacific	Randall 1985, Ralston et al. 1986, Randall et al. 1990
Prognathodes guezei (Mauge and Bauchot)	O	107–214	Hard, sand	HA	60–200	Pacific	Pyle and Chave 1994
Family Pomacanthidae							
Centropyge potteri Jordan and Metz	O	43–138	Various	LH, PB, WH	10–55	Johnston Atoll, Hawaiian Is.	Strasburg et al. 1968; Randall 1985, 1996; Randall et al. 1985
Genicanthus personatus Randall	O	46–174	Various	LH, PB, WH	24–85	Hawaiian Is.	Randall 1975, 1996; Randall and Struhsaker 1976; Severns and Fiene-Severns 1993
Holacanthus arcuatus Gray	C	43–208	Various	HA, JA	12–183	Johnston Atoll, Hawaiian Is.	Okamoto and Kanenaka 1984; Randall 1985, 1996; Randall et al. 1985
Family Pentacerotidae							
Evistias acutirostris (Schlegel)	O	89–193	Hard, sand	HA	84–183	Pacific	Hardy 1983, Okamoto and Kanenaka 1984, Severns and Fiene-Severns 1993, Randall 1996
Family Owstoniidae							
Owstonia sp.	O	349–420	Hard, holes	CS, LH, WH			Grigg et al. 1987
Family Carangidae							
Carangoides orthogrammus (Jordan and Gilbert)	R	83–190	Midwater	JA, HA	3–168	Indo-Pacific	Randall et al. 1985, 1990; Myers 1989

Species						Sources
Caranx ignobilis (Forskål)	R	80–188	Various	LH, PB, WH	1–80	Indo-Pacific	Myers 1989, Hobson and Chave 1990, Randall et al. 1990
Caranx lugubris Poey	C	61–291	Hard	JA	12–354	Circumglobal	Randall et al. 1985, 1990; Myers 1989
Caranx melampygus Cuvier	O	31–150	Sand	HA	1–230	Indo-Pacific	Ralston et al. 1986, Myers 1989, Randall et al. 1990
Decapterus macarellus (Cuvier)	R	61–107	Midwater, sand	PB	1–200	Circumglobal	Myers 1989, Randall et al. 1990, Randall 1996
Decapterus tabl Berry	O	361–416	Midwater, sand	WO	125–530	Circumglobal	Okamura and Kitajima 1984, Uchida and Tagami 1984, Randall et al. 1990
Elagatis bipinnulata Bennett	R	122–144	Midwater	JA, WH	1–150	Circumglobal	Masuda et al. 1984, Myers 1989, Randall et al. 1990
Pseudocaranx dentex (Bloch and Schneider)	O	104–150	Various	NWHI	80–200	Circumglobal	Randall 1981a, Uchida and Tagami 1984, Randall et al. 1990
Seriola dumerili (Risso)	A	50–385	Various	HA, JA	8–335	Circumglobal	Ohta 1983; Randall et al. 1985, 1990; Ralston et al. 1986; Myers 1989
Family Cirrhitidae							
Cirrhitops fasciatus (Bennett)	R	49–52	Hard	PB	Shallow	Indo-Pacific	Randall 1963, 1996; Hobson and Chave 1990
Paracirrhites arcatus (Cuvier)	R	55–57	Coral heads	PB	1–91	Indo-Pacific	Strasburg et al. 1968, Myers 1989, Randall et al. 1990, Randall 1996
Family Pomacentridae							
Chromis leucura Gilbert	O	46–122	Various	LH, PB, WO	29–118	Central Pacific	Randall and Swerdloff 1973, Masuda et al. 1984, Randall et al. 1985, Randall 1996
Chromis ovalis (Steindachner)	C	46–161	Various	LH, PB, WH	7–45	Hawaiian Is.	Randall and Swerdloff 1973, Struhsaker 1973, Hobson 1984, Randall 1996
Chromis struhsakeri Randall and Swerdloff	C	131–250	Various	LH, PB, WO	99–302	Hawaiian Is.	Struhsaker 1973, Uchida and Tagami 1984
Chromis verater Jordan and Metz	A	37–199	Various	HI, JA	10–140	Indo-Pacific	Randall and Swerdloff 1973, Struhsaker 1973, Randall et al. 1985, Ralston et al. 1986, Randall 1996
Dascyllus albisella Gill	C	28–84	Various	JA, LH, PB	1–45	Johnston Atoll, Hawaiian Is.	Randall and Swerdloff 1973; Randall 1985, 1996; Randall et al. 1985
Family Labridae							
Anampses chrysocephalus Randall	O	61–139	Various	PB	10–15	Hawaiian Is.	Okamoto and Kanenaka 1984, Severns and Fiene-Severns 1993, Randall 1996

(continued on next page)

Table 1
A Taxonomic List of Animals Photographed from HURL Submersibles (continued)

TAXON	Rank* (observed)	Depth (m) (observed)	Substrate (on/above)	Location** (observed)	Depth (m) (published)	Location (published)	Selected References (depth, location, and/or photograph)
Bodianus bilunulatus (Lacépède)	O	52–177	Various	JA, LH, WH	8–200	Indo-Pacific	Okamura 1985, Ralston et al. 1986, Randall 1996
Bodianus cylindriatus (Tanaka)	R	156–183	Hard, holes	PB	240–510	Pacific	Okamura and Kitajima 1984, Randall and Chen 1985
Bodianus sanguineus (Jordan and Evermann)	R	104–168	Various	LH, PB, WO	32–68	Hawaiian Is.	Madden 1973, Gomon and Randall 1978, Randall 1996
Bodianus vulpinus (Richardson)	R	185–190	Various	PB	6–183	Pacific	Gomon and Randall 1978, Randall 1981a, Humphreys et al. 1984
Coris ballieui Vaillant and Sauvage	R	83	Sand	PB	15–85	Hawaiian Is.	Severns and Fiene-Severns 1993, Randall 1996
Coris flavovittata (Bennett)	R	61–98	Various	LO, PB, WO	1–61	Hawaiian Is.	Strasburg et al. 1968, Randall 1996
Coris gaimard (Quoy and Gaimard)	R	61	Hard, holes	PB	1–46	Indo-Pacific	M. M. Smith and Heemstra 1986, Myers 1989, Randall et al. 1990, Randall 1996
Cymolutes lecluse (Quoy and Gaimard)	R	55–119	Sand	LH, LO, WO	5–30	Hawaiian Is.	Strasburg et al. 1968, Randall 1996
Labroides phthirophagus Randall	R	46–122	Hard	JA, HI	2–91	Johnston Atoll, Hawaiian Is.	Randall 1985, Randall et al. 1985, Strasburg et al. 1968
Oxycheilinus bimaculatus (Valenciennes)	R	57–92	Talus	PB	2–110	Indo-Pacific	Myers 1989, Randall et al. 1990, Westneat 1993
Oxycheilinus unifasciatus (Streets)	R	61–76	Various	LH	10–165	Pacific	Myers 1989, Randall et al. 1990, Westneat 1993
Polylepion russelli (Gomon and Randall)	R	92–318	Various	JA, LH	240–280	Pacific	Gomon and Randall 1978, Randall et al. 1985
Pseudocheilinus evanidus Jordan and Evermann	R	52–92	Talus	LH, PB	6–42	Indo-Pacific	Masuda et al. 1984, Myers 1989, Randall 1996
Pseudojuloides cerasinus (Snyder)	C	49–104	Talus	PB, WH	2–61	Indo-Pacific	Randall 1985, 1996; Myers 1989
Suezichthys notatus (Kamohara)	R	122–272	Various	HA	119–204	Pacific	Randall and Kotthaus 1977, Russell 1985
Family Percophidae							
Bembrops filifera Gilbert	R	300–365	Sand	LA, LO	226–440	Central Pacific	Struhsaker 1973, Humphreys et al. 1984, Okamura 1985
Chrionema chryseres Gilbert	O	303–455	Various	LA, LH, LO, WH, WO	234–500	Pacific	Struhsaker 1973, Iwamoto and Staiger 1976, Masuda et al. 1984

Species							References
Chrionema squamiceps Gilbert	O	306–520	Sand	HA	174–600	Johnston Atoll, Hawaiian Is.	Struhsaker 1973, Iwamoto and Staiger 1976, Randall et al. 1985
Family Pinguipedidae							
Parapercis roseoviridis (Gilbert)	O	150–300	Sand	HI, JA	183–300	Central Pacific	Struhsaker 1973, Masuda et al. 1984, Randall et al. 1985
Parapercis schauinslandi (Steindachner)	C	49–141	Sand	LH, LO, PB, WO	15–170	Indo-Pacific	Struhsaker 1973, Masuda et al. 1984, Randall et al. 1990, Severns and Fiene-Severns 1993, Randall 1996
Family Draconettidae							
Centrodraco rubellus Frick, Chave and Suzumoto	R	367	Sand	LO	280–363	Pacific	Fricke 1992
Draconetta xenica Jordan and Fowler	R	284–367	Sand	LO, WO	180–230	Indo-Pacific	Nakabo 1982, Parin 1982
Family Zanclidae							
Zanclus cornutus (Linnaeus)	O	31–147	Various	JA, LH, LO, PB, WH, WO	2–183	Indo-Pacific	Strasburg et al. 1968, Myers 1989, Randall et al. 1990, Randall 1996
Family Acanthuridae							
Acanthurus dussumieri Valenciennes	O	31–118	Various	JA, LH, PB	9–131	Indo-Pacific	Randall et al. 1985, Myers 1989, Randall 1996
Acanthurus nigricans (Linnaeus)	R	92–101	Various	NWHI, PB	4–67	Indo-Pacific	Myers 1989, Randall 1996
Acanthurus olivaceus Forster	R	44–83	Hard	LH, LO, PB, WO	9–62	Pacific	Myers 1989, Randall et al. 1990, Randall 1996
Acanthurus thompsoni (Fowler)	R	46–119	Sand	JA	4–75	Indo-Pacific	Randall et al. 1985, Myers 1989, Randall 1996
Acanthurus xanthopterus Valenciennes	O	31–120	Various	LH, PB, WH	5–90	Indo-Pacific	Myers 1989, Randall et al. 1990, Randall 1996
Ctenochaetus strigosus (Bennett)	O	15–113	Various	LH, WH	1–46	Indo-Pacific	Strasburg et al. 1968, Myers 1989, Randall et al. 1990
Naso brevirostris (Valenciennes)	R	54–122	Various	JA, PB	4–46	Indo-Pacific	Strasburg et al. 1968, Myers 1989, Randall et al. 1990
Naso hexacanthus (Bleeker)	C	46–229	Various	HA	6–150	Indo-Pacific	Okamura 1985, Ralston et al. 1986, Myers 1989, Randall et al. 1990
Naso lituratus (Bloch and Schneider)	R	58–76	Various	LH, NWHI, PB	4–90	Indo-Pacific	Randall 1985, 1996; Myers 1989
Naso maculatus Randall and Struhsaker	O	76–120	Various	PB	43–220	Pacific	Randall and Struhsaker 1981, Okamura 1985, Randall 1996
Zebrasoma flavescens (Bennett)	O	15–81	Various	JA, LH, WH	1–50	Pacific	Masuda et al. 1984, Randall 1985, Randall et al. 1990
Family Gempylidae							
Rexea nakamurai Parin	R	378–420	Hard, sand	LA, WO	340–374	Pacific	Struhsaker 1973, Parin 1989

(continued on next page)

Table 1
A Taxonomic List of Animals Photographed from HURL Submersibles (continued)

TAXON	Rank* (observed)	Depth (m) (observed)	Substrate (on/above)	Location** (observed)	Depth (m) (published)	Location (published)	Selected References (depth, location, and/or photograph)
Family Pleuronectidae							
Poecilopsetta hawaiiensis Gilbert	R	245–367	Sand	LA, LO, PB, WM, WO	183–422	Hawaiian Is.	Struhsaker 1973
Family Bothidae							
Chascanopsetta prorigera Gilbert	R	343–367	Sand	HI	260–450	Hawaiian Is.	Struhsaker 1973, Uchida and Tagami 1984
Parabothus coarctatus (Gilbert)	R	180–398	Sand	LO, WH, WO	180–400	Pacific	Uchida and Tagami 1984; A. M. Suzumoto, pers. comm., 1990
Taeniopsetta radula Gilbert	R	180–275	Sand	LO, PB, WO	65–116	Hawaiian Is.	Struhsaker 1973
Family Triacanthodidae							
Hollardia goslinei Tyler	O	275–515	Various	CS, LA, LH, LO, WH, WO	335–365	Johnston Atoll, Hawaiian Is.	Clarke 1972, Humphreys et al. 1984, Randall et al. 1985
Family Balistidae							
Aluterus scriptus (Osbeck)	R	55–120	Various	LO, PB, WO	2–116	Circumglobal	Strasburg et al. 1968, Fielding and Robinson 1987, Myers 1989
Balistes polylepis Steindachner	R	47–60	Hard, sand	PB	10–50	Pacific	Gosline and Brock 1960, Randall 1996
Melichthys niger (Bloch)	R	40–52	Various	PB	4–75	Circumglobal	Randall and Klausewitz 1973, Myers 1989, Randall 1996
Melichthys vidua (Solander)	R	61–80	Sand, hard	PB	4–145	Indo-Pacific	Randall and Klausewitz 1973, Myers 1989, Randall 1996
Pervagor spilosoma (Lay and Bennett)	A	52–138	Sand, talus	LH, LO, PB, WO	6–150	Johnston Atoll, Hawaiian Is.	Randall et al. 1985, Hobson 1984, Hutchins 1986, Fielding and Robinson 1987
Sufflamen fraenatus (Latreille)	O	31–183	Hard, holes	JA, LH, PB, WH	18–170	Indo-Pacific	Randall 1985, Ralston et al. 1986, Myers 1989
Xanthichthys auromarginatus (Bennett)	O	31–161	Hard, holes	HI, JA	12–147	Indo-Pacific	Randall et al. 1978, 1985; Randall 1996
Xanthichthys mento (Jordan and Gilbert)	R	61–131	Hard, holes	PB	6–40	Pacific	Randall et al. 1978, Masuda et al. 1984, Severns and Fiene-Severns 1993, Randall 1996
Family Tetraodontidae							
Arothron hispidus (Linnaeus)	R	31–73	Sand	LO, WO	1–99	Indo-Pacific	Struhsaker 1973, Randall et al. 1990, Randall 1996
Arothron meleagris (Bloch and Schneider)	R	67–73	Sand	LO, WH	1–14	Indo-Pacific	Myers 1989, Randall 1996

Species		Range	Substrate	Habitat	Depth (m)	Distribution	References
Canthigaster coronata (Vaillant and Sauvage)	O	61–92	Various	LO, PB, WO	11–120	Indo-Pacific	Allen and Randall 1977, Myers 1989, Randall et al. 1990
Canthigaster epilampra (Jenkens)	R	85–119	Sand	PB	9–36	Indo-Pacific	Myers 1989, Randall 1996
Canthigaster inframacula Allen and Randall	R	165–168	Various	JA	124–274	Pacific	Allen and Randall 1977, Matsuura and Yoshino 1984, Randall et al. 1985
Canthigaster jactator (Jenkens)	R	61–89	Various	PB	1–20	Hawaiian Is.	Randall 1985, 1996
Canthigaster rivulata (Schlegel)	R	135–278	Various	LO, PB, WO	95–357	Indo-Pacific	Madden 1973, Struhsaker 1973, Allen and Randall 1977, Okamura 1985, Randall 1996
Sphoeroides pachygaster (Müller and Troschel)	R	248–367	Sand	WH	80–410	Circumglobal	Struhsaker 1973, Hardy 1981, M. M. Smith and Heemstra 1986
Family Diodontidae							
Chilomycterus reticulatus (Linnaeus)	R	61–141	Hard, holes	LO, PB, WO	1–123	Circumglobal	Struhsaker 1973, Masuda et al. 1984, Randall et al. 1990, Randall 1996

References

Agassiz, A., and H. L. Clark. 1908–1909. Hawaiian and other Pacific Echini. *Memoirs of the Museum of Comparative Zoology, Harvard* 34, nos. 1–3:1–383.

Allen, G. R. 1979. *Butterfly and angelfishes of the world.* Vol. 2, *Atlantic Ocean, Caribbean Sea, Red Sea, Indo-Pacific.* Mell: Mergus/Hans Baensch.

———. 1985. Snappers of the world (F.A.O. Species Catalogue). *F.A.O. Fisheries Synopsis,* no. 125, vol. 6, pt. 1, pp. 1–208.

Allen, G. R., and J. E. Randall. 1977. Review of the sharpnose pufferfishes (Subfamily Canthigasterinae) of the Indo-Pacific. *Records of the Australian Museum* 30, no. 17:475–517.

Allen, G. R., and F. H. Talbot. 1985. Review of the snappers of the genus *Lutjanus* (Pisces: Lutjanidae) from the Indo-Pacific with a description of a new species. I*ndo-Pacific Fishes,* no. 11:1–87.

Anderson, W. D., Jr. 1970. Revision of the genus *Symphysanodon* (Pisces: Lutjanidae) with descriptions of four new species. *United States Fishery Bulletin* 68, no. 2:325–346.

———. 1981. A new species of Indo-West Pacific *Etelis* (Pisces: Lutjanidae), with comments on other species of the genus. *Copeia,* no. 4:820–825.

Baba, K. 1974. *Munida brucei* sp. nov., a new galatheid (Decapoda, Anomura) from the east coast of Africa. *Annals of Zoology,* Japan 47:55–60.

———. 1981. A new galatheid crustacean (Decapoda, Anomura) from the Hawaiian Islands. *Journal of Crustacean Biology* 1, no. 2:288–292.

Bailey-Brock, J. H. 1972. Deepwater tube worms (Polychaeta, Serpulidae) from the Hawaiian Islands. *Pacific Science* 26, no. 4:405–408.

Bate, C. S. 1888. Report on the Crustacea Macrura collected by the H.M.S. *Challenger.* In *Report of the scientific results of the voyage of the H.M.S. Challenger during the years 1873–1876* (Zoology section 5), ed. C. W. Thompson and J. Murray, 24:1–942. London: Eyre & Spottiswoode.

Bayer, F. M. 1952. Descriptions and redescriptions of the Hawaiian octocorals collected by the United States Fish Commission steamer *Albatross:* 1. Alcyonacea, Stolonifera and Telestacea. *Pacific Science* 6:126–136.

———. 1956. Descriptions and redescriptions of the Hawaiian octocorals collected by the United States Fish Commission steamer *Albatross:* 2. Gorgonacea: Scleraxonia. *Pacific Science* 10:67–95.

———. 1981. Key to the genera of Octocorallia exclusive to Pennatulacea (Coelenterata: Anthozoa), with diagnoses of new taxa. *Proceedings of the Biological Society of Washington* 94, no. 3:902–947.

Bayer, F. M., and J. Stefani. 1988. A new species of *Chrysogorgia* (Octocorallia: Gorgonacea) from New Caledonia, with descriptions of some other species from the western Pacific. *Proceedings of the Biological Society of Washington* 101, no. 2:257–279.

Berry, S. S. 1914. The cephalopods of the Hawaiian Islands. *Washington Bureau of Fisheries Bulletin* 32:255–362.

Billett, D. S. M., and B. Hansen. 1982. Abyssal aggregations of *Kolga hyalina* Danielssen and Koren (Echinodermata: Holothurioidea) in the northeast Atlantic Ocean: A preliminary report. *Deep-Sea Research* 29, no. 7A:799–818.

Borets, L. A. 1986. Ichthyofauna of the northwestern and Hawaiian submarine ranges. *Journal of Ichthyology* 26, no. 3:1–13.

Brock, V. E., and T. C. Chamberlain. 1968. A geological and ecological reconnaissance off western Oahu, Hawaii, principally by means of the research submarine *Asherah*. *Pacific Science* 22, no. 3:373–394.

Brook, G. 1889. Report on the antipatharians collected by the H.M.S. *Challenger*. In *Report of the scientific results of the voyage of the H.M.S. Challenger during the years 1873–1876* (Zoology section 5), ed. C. W. Thompson and J. Murray, 32, no. 1:1–222. London: Eyre & Spottiswoode.

Bryan, W. B., and R. S. Stephens. 1993. Coastal bench formation at Hanauma Bay, Oahu, Hawaii. *Bulletin of the Geological Society of America* 105, no. 3:377–386.

Burkenroad, M. D. 1936. The Aristaeinae, Solenocerinae and pelagic Penaeinae of the Bingham Oceanographic Collection. *Bulletin of the Bingham Oceanographic Collection* 5, no. 2:1–151.

Burukovskii, R. N. 1980. Some biological aspects of the shrimp *Plesiopenaeus edwardsianus* in the southeast Atlantic. *Biologie Morya* 6:21–26.

Busahkin, S. V. 1982. Systematics and distribution of the family Berycidae (Osteichthyes) in the world ocean. *Journal of Ichthyology* 22, no. 6:1–21.

Cairns, S. D. 1984. New records of ahermatypic corals (Scleractinia) from the Hawaiian and Line Islands. *Occasional Papers of the Bernice P. Bishop Museum* 24, no. 10:1–30.

Carlgren, O. 1949. A survey of the Ptychodactiaria, Corallimorpharia and Actiniaria. *Kungliga Svenska Vetenkapsakadamie Handlung*, ser. 4, 1, no. 1:1–121.

Caruso, J. H. 1981. The systematics and distribution of the lophiid anglerfishes: 1. A revision of the genus *Lophiodes* with the description of two new species. *Copeia*, no. 3:522–549.

———. 1985. The systematics and distribution of the lophiid anglerfishes: 3. Intergeneric relationships. *Copeia*, no. 4:870–875.

———. 1989. Systematics and distribution of the Atlantic chaunacid anglerfishes (Pisces: Lophiiformes). *Copeia*, no. 1:153–165.

Caruso, J. H., and H. R. Bullis. 1976. A review of the lophiid angler fish genus *Sladenia* with a description of a new species from the Caribbean Sea. *Bulletin of Marine Science* 26, no. 1:59–64.

Chace, F. A. 1985. The caridian shrimp (Crustacea: Decapoda) of the Albatross Philippine expedition, 1907–1910: 3. Families Thalassocarididae and Pandalidae. *Smithsonian Contributions in Zoology* 411:1–143.

Chave, E. H. 1978. Ecology of six species of Hawaiian cardinalfishes. *Pacific Science* 32, no. 3:245–270.

Chave, E. H., and A. T. Jones. 1991. Deep-water megafauna of the Kohala and Haleakala slopes, Alenuihaha Channel, Hawaii. *Deep-Sea Research* 38, no. 7:781–803.

Chave, E. H., and E. A. Kay. 1985. Reef and shore communities. In *Atlas of Hawaii*, ed. R. W. Armstrong, 84–86. Honolulu: University of Hawai'i Press.

Chave, E. H., and B. C. Mundy. 1994. Deep-sea benthic fishes of the Hawaiian Archipelago, Cross Seamount and Johnston Atoll. *Pacific Science* 48:367–409.

Chiswell, S., E. Firing, D. Karl, R. Lukas, and C. Winn. 1990. Hawaii ocean time series data report 1, 1988–1989. *University of Hawaii SOEST Technical Report*, no. 1:1–269.

Clague, D. A., and G. B. Dalrymple. 1987. The Hawaiian-Emperor volcanic chain: 1. Geologic evolution. *United States Geological Survey Professional Paper*, no. 1350:5–54.

Clark, A. H. 1908. Descriptions of new species of crinoids chiefly from the collections made by the United States Fisheries Steamer *Albatross* at the Hawaiian Islands in 1902; with remarks on the classification of the Comatulida. *Proceedings of the United States National Museum* 34:209–239.

———. 1949. Ophiuroidea of the Hawaiian Islands. *Bernice P. Bishop Museum Bulletin* 195:1–133.

Clark, A. M. 1989. An index of names of recent Asteroidea: 1. Paxillosida and Notomyotida. In *Echinoderm studies*, ed. M. Jangoux and J. M. Lawrence, 3:225–348. Rotterdam: A. A. Balkema.

———. 1993. An index of names of recent Asteroidea: 2. Valvatida. In *Echinoderm studies*, ed. M. Jangoux and J. M. Lawrence, 4:187–366. Rotterdam: A. A. Balkema.

Clark, A. M., and F. W. Rowe. 1971. *Monograph of the shallow-water Indo-West Pacific echinoderms.* London: British Museum of Natural History. (238 pp.)

Clark, H. L. 1914. Hawaiian and other Pacific Echini. *Memoirs of the Museum of Comparative Zoology, Harvard* 46, no. 1:1–78.

———. 1917. Hawaiian and other Pacific Echini. *Memoirs of the Museum of Comparative Zoology, Harvard* 46, no. 2:79–283.

Clarke, T. A. 1972. Collections and submarine observations of deep benthic fishes and decapod crustacea in Hawaii. *Pacific Science* 26, no. 3:310–317.

Compagno, L. J. V. 1984. Sharks of the world (F.A.O. Species Catalogue). *F.A.O. Fishery Synopsis,* no. 125, vol. 4, pt. 1, pp. 1–249; pt. 2, pp. 1–655.

Davidson, T. 1880. Report on the Brachiopoda collected by the H.M.S. *Challenger.* In *Report of the scientific results of the voyage of the H.M.S. Challenger during the years 1873–1876* (Zoology section 5), ed. C. W. Thompson and J. Murray, 1, no. 1:1–67. London: Eyre & Spottiswoode.

Dawson, E. W., and J. C. Yaldwyn. 1985. *Lithodes nintokuae* Sakai: A deep-water king crab newly recorded from Hawaii. *Pacific Science* 39, no. 1:16–23.

DeCarlo, E. H., G. M. McMurtry, and H. Yeh. 1983. Geochemistry of the hydrothermal deposits from Loihi submarine volcano, Hawaii. *Earth and Planetary Science Letters* 66:438–449.

Debelius, H. 1984. *Armoured knights of the sea.* Essen: Kernen Verlag. (120 pp.)

DeLaubenfels, M. W. 1963. Part E. Porifera. In *Treatise on invertebrate paleontology,* ed. R. C. Moore, 21–122. New York: Geological Society of America.

DeMan, J. G. 1920. Decapoda of the Siboga expedition. In *Siboga Expeditie, Oost-Indie, 1899–1900,* ed. M. Weber, no. 39A, pt. 4:1–318. Leiden: E. J. Brill.

Devaney, D. M. 1977. Class Anthozoa. In *Reef and shore fauna of Hawaii,* vol. 1, ed. D. M. Devaney and L. G. Eldredge, 119–129. Bernice P. Bishop Museum Special Publication 64. Honolulu.

———. 1987. Echinodermata other than Holothuroidea. In *The natural history of Enewetak Atoll,* pt. 2, ed. D. M. Devaney, E. S. Reese, B. L. Burch, and P. Helfrich, 277–285. Washington, D.C.: U.S. Department of Energy.

Dunn, D. F. 1982. Cnidaria. In *Synopsis and classification of living organisms,* pt. 1, ed. S. P. Parker, 669–706. New York: McGraw-Hill.

Dunn, D. F., and J. Backus. 1977. Re-description and ecology of *Liponema brevicornis. Astarte* 10:77–85.

Dunn, D. F., D. M. Devaney, and B. Roth. 1980. *Stylobates,* a shell forming anemone (Coelenterata, Anthozoa, Actiniidae). *Pacific Science* 34, no. 4:379–388.

Eaton, J. P. 1962. Crustal structure and volcanism in Hawaii. In *The crust of the Pacific Basin,* ed. G. A. Macdonald and H. Kuno, 13–29. American Geophysical Union Monograph 6. Washington, D.C.

Eckelbarger, K. I., C. M. Young, and J. L. Cameron. 1989. Ultrastructure and development of dimorphic sperm in the abyssal echinoid *Phrissocystis multispina* (Echinodermata: Echinoidea): Implications for deep sea reproductive biology. *Biological Bulletin* 176:257–271.

Edmondson, H. E. 1952. Additional central Pacific crustaceans. *Occasional Papers of the Bernice P. Bishop Museum* 21, no. 6:67–86.

Eldredge, L. G., and S. E. Miller. 1995. Records of the Hawaii Biological Survey for 1994: How many species are there in Hawaii? *Occasional Papers of the Bernice P. Bishop Museum,* no. 41:1–8.

Ely, C. A. 1942. Shallow-water Asteroidea and Ophiuroidea of Hawaii. *Bernice P. Bishop Museum Bulletin* 176:9–30.

Emson, R. H., and C. M. Young. 1994. Feeding mechanisms of the brisingid starfish *Novodinia antillensis. Marine Biology* 118:433–442.

Eschmeyer, W. N., and J. E. Randall. 1975. The scorpaenid fishes of the Hawaiian Islands, including new species and new records (Pisces: Scorpaenidae). *Proceedings of the California Academy of Sciences,* 4th ser., 40, no. 11:265–334.

Fielding, A., and E. Robinson. 1987. *Underwater guide to Hawai'i.* Honolulu: University of Hawai'i Press.

Fisher, W. K. 1906. The starfishes of the Hawaiian Islands. In *Aquatic resources of the Hawaiian Islands,* ed. D. S. Jordan. *Bulletin of the United States Fish Commission for 1903* 23, pt. 3:989–1190.

———. 1907. The holothurians of the Hawaiian Islands. *Proceedings of the United States National Museum* 32, no. 1555:637–744.

———. 1925. Seastars of tropical central Pacific. *Bernice P. Bishop Museum Bulletin* 27:63–88.

Fornari, D. J., M. O. Garcia, R. C. Tyce, and D. G. Gallo. 1988. Morphology and structure of Loihi Seamount based on Seabeam sonar mapping. *Journal of Geological Research* 93:15227–15238.

Fornari, D. J., A. Malahoff, and B. C. Heezen. 1978. Volcanic structure of the crest of the Puna Ridge, Hawaii: Geophysical implications of submarine volcanic terrain. *Bulletin of the Geological Society of America* 89:605–616.

Foster, M. W. 1982. Brachiopoda. In *Synopsis and classification of living organisms,* pt. 2, ed. S. Parker, 773–780. New York: McGraw-Hill.

France, S. C., P. E. Rosel, J. E. Agenbroad, L. S. Mullineaux, and T. D. Kocher. 1996. DNA sequence variation of mitochondrial large-subunit rRNA provides support for a two-subclass organization of the Anthozoa (Cnidaria). *Molecular Marine Biology and Biotechnology* 4, no. 1:15–28.

Fraser, T. H. 1972. Comparative osteology of the shallow water cardinalfishes (Perciformes: Apogonidae) with reference to the systematics and evolution of the family. *Bulletin of the J. L. B. Smith Institute of Ichthyology,* no. 34:1–105.

Freitas, A. J. 1985. The Penaeoidea of southeast Africa: 2. The families Aristeidae and Solenoceridae. *South African Association of Marine Biological Research, Oceanographic Research Institute Investigations Report* 57:1–69.

Fricke, R. 1992. Revision of the family Draconettidae (Teleostei), with descriptions of two new species and a new subspecies. *Journal of Natural History* 26:165–195.

Genin, A., P. K. Dayton, P. F. Lonsdale, and F. N. Spiess. 1986. Corals on seamount peaks provide evidence of current acceleration over deep-sea topography. *Nature* 322, no. 6074:59–61.

Gilbert, C. H. 1905. The deep-sea fishes. In *Aquatic resources of the Hawaiian Islands,* ed. D. S. Jordan. *Bulletin of the United States Fish Commission for 1903* 23, pt. 2:577–713.

Gilbert, C. H., and F. Cramer. 1892. Report on the fishes dredged in deep water near the Hawaiian Islands with descriptions and figures of twenty-three new species. *Proceedings of the United States National Museum* 19, no. 1114:403–435.

Golovan, A. A., and N. P. Pakhorukov. 1987. Distribution and behavior of fishes on Naska and Sala y Gomez submarine ridges. *Journal of Ichthyology* 3:369–376.

Gomon, M. F., and J. E. Randall. 1978. Review of the Hawaiian fishes of the labrid tribe Bodianini. *Bulletin of Marine Science* 28, no. 1:32–48.

Gon, O. 1983. *Paratrachichthys heptalepis,* a new roughie (Pisces Trachichthyidae) from the Hawaiian Islands. *Pacific Science* 37, no. 3:293–299.

———. 1985. Two new species of the deep-sea cardinalfish genus *Epigonus* (Perciformes, Apogonidae) from the Hawaiian Islands, with a key to the Hawaiian species. *Pacific Science* 39, no. 2:221–229.

———. 1987. New records of three fish species from Hawaii. *Japanese Journal of Ichthyology* 34, no. 1:100–104.

Goodbody, I. 1982. Tunicata. In *Synopsis and classification of living organisms,* pt. 2, ed. S. Parker, 823–829. New York: McGraw-Hill.

Gooding, R. M., J. J. Polovina, and M. D. Dailey. 1988. Observations of deepwater shrimp, *Heterocarpus ensifer,* from a submersible off the Island of Hawaii. *Marine Fisheries Review* 50, no. 1:32–39.

Gosline, W. A. 1953. Hawaiian shallow-water fishes of the family Brotulidae, with a description of a new genus and notes on brotulid anatomy. *Copeia,* no. 4:215–225.

Gosline, W. A., and V. E. Brock. 1960. *Handbook of Hawaiian fishes.* Honolulu: University of Hawai'i Press. (372 pp.)

Goy, J. W., and D. M. Devaney. 1980. *Stenopus pyrsonotus,* a new species of stenopodidean shrimp from the Indo-West Pacific region (Crustacea Decapoda). *Proceedings of the Biological Society of Washington* 93, no. 3:781–796.

Gray, J. E. 1867. Notes on the arrangement of sponges, with the description of some new genera. *Proceedings of the Zoological Society of London* 27:492–558.

———. 1870. Notes on anchoring sponges (in a letter to Mr. Moore). *Annals and Magazine of Natural History* 6, no. 6:309–312.

Greenfield, D. W. 1974. A revision of the squirrelfish genus *Myripristis* Cuvier (Pisces: Holocentridae). *Bulletin of the Natural History Museum, Los Angeles County,* no. 19:1–54.

Grigg, R. W. 1977. *Hawaii's precious corals.* Honolulu: Island Heritage Press. (27 pp.)

———. 1988. Recruitment limitation of a deep benthic hard-bottom octocoral population in the Hawaiian Islands. *Marine Ecology,* progress ser., 45:121–126.

Grigg, R. W., and F. M. Bayer. 1976. Present knowledge of the systematics and zoogeography of the order Gorgonacea in Hawaii. *Pacific Science* 30, no. 2:167–175.

Grigg, R. W., and L. G. Eldredge. 1975. The commercial potential of precious corals in Micronesia: 1. The Marianas Islands. *University of Guam Technical Report* 18:1–16.

Grigg, R. W., A. Malahoff, E. H. Chave, and J. Landahl. 1987. Seamount benthic ecology and potential environmental impact from manganese crust mining in Hawaii. In *Seamounts, islands, and atolls,* ed. B. H. Keating, P. Fryer, R. Batiza, and G. W. Boehlert, 379–390. American Geophysical Union Monograph 43. Washington, D.C.

Grigg, R. W., and D. Opresko. 1977. Order Antipatharia (black corals). In *Reef and shore fauna of Hawaii,* vol. 1, ed. D. M. Devaney and L. G. Eldredge, 242–261. Bernice P. Bishop Museum Special Publication 64. Honolulu.

Hansen, B. 1975. Systematics and biology of the deep-sea holothurians. *Galathea Report* 13:1–262.

Hardy, G. S. 1981. New records of pufferfishes (Family Tetraodontidae) from Australia and New Zealand with notes on *Sphoeroides pachygaster* (Muller and Troschel) and *Lagocephalus sceleratus* (Gmelin). *Records of the National Museum, New Zealand* 1, no. 20:311–316.

———. 1983. A revision of the fishes of the family Pentacerotidae (Perciformes). *New Zealand Journal of Zoology* 10:177–220.

Hartman, W. D. 1982. Porifera. In *Synopsis and classification of living organisms,* pt. 1, ed. S. Parker, 640–666. New York: McGraw-Hill.

Heemstra, P. C. 1980. A revision of the zeid fishes (Zeiformes: Zeidae) of South Africa. *Bulletin of the J. L. B. Smith Institute of Ichthyology,* no. 41:1–18.

Heemstra, P. C., and J. E. Randall. 1977. A revision of the Emmelichthyidae (Pisces: Perciformes). *Australian Journal of Marine and Freshwater Research* 28:361–396.

Herald, E. S., and J. E. Randall. 1972. Five new Indo-Pacific pipefishes. *Proceedings of the California Academy of Sciences* 39, no. 11:121–140.

Hertwig, R. 1888. Report on the Actiniaria collected by the H.M.S. *Challenger.* In *Report of the scientific results of the voyage of the H.M.S. Challenger during the years 1873–1876* (Zoology section 5), ed. C. W. Thompson and J. Murray, 26, no. 3:1–54. London: Eyre & Spottiswoode.

Hiramoto, K. 1974. Biological notes on an anomuran crab, *Lithodes longispinna* Sakai. *Researches on Crustacea* 6:17–24.

Hobson, E. S. 1984. The structure of reef fish communities in the Hawaiian Archipelago. In *Proceedings of Research Investigations in the Northwestern Hawaiian Islands* (University of Hawai'i–Seagrant MR-84-01), 101–122. Honolulu.

Hobson, E. S., and E. H. Chave. 1990. *Hawaiian reef animals.* Honolulu: University of Hawai'i Press. (398 pp.)

Holthuis, L. B. 1955. The recent genera of caridean and stenopodidean shrimps (Class Crustacea, Order Decapoda, Supersection Natantia) with keys for their determination. *Zoologische Verhandlung* 26:1–165.

Hoover, J. P. 1993. *Hawaii's fishes.* Honolulu: Mutual Publishers. (178 pp.)

Howes, G. J., and O. A. Crimmen. 1990. A review of the Bathygadidae (Teleostei: Gadiformes). *Bulletin of the British Museum of Natural History (Zoology)* 56, no. 2:155–203.

Humphreys, R. L., Jr., D. Tagami, and M. P. Seki. 1984. Seamount fishery resources within the southern Emperors–Northern Hawaiian Ridge area. In *Proceedings of research investigations in the Northwestern Hawaiian Islands* (University of Hawai'i–Seagrant MR-84-01), 283–327. Honolulu.

Hutchins, J. B. 1986. Review of the monacanthid fish genus *Pervagor* with descriptions of two new species. *Indo-Pacific Fishes,* no. 12:1–35.

Ijima, I. 1901. Studies on the Hexactinellida. *Journal of the College of Sciences, Imperial University, Tokyo* 15, no. 1:1–299.

———. 1902. Studies on the Hexactinellida: 2. The genera *Corbitella* and *Heterotella*. *Journal of the College of Sciences, Imperial University, Tokyo* 17, no. 9:1–34.

Iwamoto, T., and J. C. Staiger. 1976. Percophidid fishes of the genus *Chrionema* Gilbert. *Bulletin of Marine Science* 26, no. 4:488–498.

Jackson, E. S., E. A. Silver, and G. B. Dalrymple. 1972. Hawaiian-Emperor chain and its relation to Cenozoic circum-Pacific tectonics. *Bulletin of the Geological Society of America* 83:601–617.

James, N. P. 1983. Reef environment. In *Carbonate depositional environments*, ed. P. A. Scholle, D. G. Bebout, and C. H. Moore, 347–440. Memoirs of the American Association of Petroleum Geologists 33. Tulsa.

Johnson, R. H. 1984. *Sharks of Polynesia*. Singapore: Editions du Pacifique. (170 pp.)

Jones, A. T. 1993. Review of the chronology of marine terraces in the Hawaiian Archipelago. *Quarterly Science Reviews* 12:811–823.

Jones, A. T., and K. J. Sulak. 1990. First central Pacific plate and Hawaiian record of the deep-sea tripod fish *Bathypterois grallator* (Pisces: Chlorophthalmidae) in Hawaiian waters. *Pacific Science* 44:254–257.

Kanayama, T. 1981. Scorpaenid fishes from the Emperor seamount chain. *Research Institute of North Pacific Fisheries, Hokkaido University, Special Volume*, 119–129.

Karl, D. M., G. M. McMurtry, A. Malahoff, and M. O. Garcia. 1988. Loihi Seamount, Hawaii: A mid-plate volcano with a distinctive hydrothermal system. *Nature* 335, no. 6190:1655–1673.

Kaufmann, R. S., W. W. Wakefield, and A. Genin. 1989. Distribution of epibenthic megafauna and lebensspuren on two central North Pacific seamounts. *Deep-Sea Research* 36, no. 12:1863–1896.

Kay, E. A. 1979. Hawaiian marine shells. In *Reef and shore fauna of Hawaii*, vol. 4, ed. D. M. Devaney and L. G. Eldredge, 1–653. Bernice P. Bishop Museum Special Publication 64. Honolulu.

Keating, B. H. 1987. Structural failure and drowning of Johnston Island (central Pacific Basin). In *Seamounts, islands, and atolls*, ed. B. H. Keating, P. Fryer, R. Batiza, and G. W. Boehlert, 49–50. American Geophysical Union Monograph 43. Washington, D.C.

Kensley, B. 1981. On the zoogeography of Southern African decapod crustacea, with a distributional checklist. *Smithsonian Contributions in Zoology* 338:1–64.

Klein, F. W. 1982. Earthquakes at Loihi submarine volcano and the Hawaiian hotspot. *Journal of Geophysical Research* 87, no. B9:7719–7726.

Kolliker, A. 1880. Report on the Pennatulida collected by the H.M.S. *Challenger*. In *Report of the scientific results of the voyage of the H.M.S. Challenger during the years 1873–1876* (Zoology section 5), ed. C. W. Thompson and J. Murray, 1, no. 2:1–41. London: Eyre & Spottiswoode.

Kotlyar, A. N. 1993. A new species of the genus *Polymixia* (Polymixiidae, Beryciformes) from the Kyushu-Palau submarine ridge and notes on other members of the genus. *Journal of Ichthyology* 33, no. 3:30–49.

Kukenthal, W. 1924. Gorgonaria. *Das Tierreich* 47:1–478.

Macdonald, G. A., A. T. Abbott, and F. L. Peterson. 1986. *Volcanos in the sea—the geology of Hawaii*. Honolulu: University of Hawaiʻi Press. (517 pp.)

Macurda, D. B., and D. L. Meyer. 1976. The identification and interpretation of stalked crinoids (Echinodermata) from deep-water photographs. *Bulletin of Marine Science* 262:205–215.

Madden, W. D. 1973. The collection of live fishes from a salvaged vessel. *Copeia*, no. 1:144–145.

Malahoff, A. 1987. Geology of the summit of Loihi submarine volcano. In *Volcanism in Hawaii* (U.S. Geological Survey Professional Paper 1350), ed. R. W. Decker, T. L. Wright, and P. H. Stauffer, 1, chap. 6:133–144. Denver.

———. 1992. Unstable geological processes, Loihi submarine volcano. *Eos* 73, no. 43:513.

Malahoff, A., R. Grigg, D. Vonderhaar, K. Kelly, and A. Arquit. 1985. Mass wasting and manganese crust growth on Cross Seamount, Hawaii. *Eos* 66, no. 46:1083.

Malahoff, A., and F. McCoy. 1967. The geologic structure of the Puna submarine ridge. *Journal of Geophysical Research* 72:541–548.

Malahoff, A., G. M. McMurtry, J. C. Wiltshire, and H. W. Yeh. 1982. Geology and chemistry of hydrothermal deposits from active submarine volcano Loihi, Hawaii. *Nature* 298, no. 5871:234–239.

Maluf, L. Y. 1988. Composition and distribution of the central Eastern Pacific echinoderms. *Los Angeles County Natural History Museum Technical Report* 2:1–242.

Manning, T. B. 1967. Reviews of the genus *Odontodactylus* (Crustacea: Stomatopoda). *Proceedings of the United States National Museum* 123, no. 3606:1–35.

Maragos, J. E. 1977. Order Scleractinia, stony corals. In *Reef and shore fauna of Hawaii*, vol. 1, ed. D. M. Devaney and L. G. Eldredge, 158–241. Bernice P. Bishop Museum Special Publication 64. Honolulu.

Maragos, J. E., and P. L. Jokiel. 1986. Reef corals of Johnston Atoll: One of the world's most isolated reefs. *Coral Reefs* 4:141–150.

Masuda, H., K. Amaoka, C. Araga, T. Uyeno, and T. Yoshino, eds. 1984. *The fishes of the Japanese Archipelago*. Tokyo: Tokai University Press. (437 pp.)

Matsuura, K., and T. Yoshino. 1984. Records of three tetraodontid fishes from Japan. *Japanese Journal of Ichthyology* 31, no. 3:331–334.

Mayer, G. F. 1974. A revision of the cardinalfish genus *Epigonus* (Perciformes, Apogonidae), with descriptions of two new species. *Bulletin of the Museum of Comparative Zoology* 146, no. 3:147–203.

McCosker, J. E. 1979. The snake eels (Pisces, Ophichthidae) of the Hawaiian Islands, with a description of two new species. *Proceedings of the California Academy of Sciences* 42, no. 2:57–67.

McDowell, S. B. 1973. Family Halosauridae. In *Fishes of the western North Atlantic*, ed. D. M. Cohen et al. *Memoirs of the Sears Foundation for Marine Research* (Yale University) 1, pt. 6:32–123.

McLaughlin, P. A., and J. H. Bailey-Brock. 1975. A new Hawaiian hermit crab of the genus *Trizopagurus* (Crustacea, Decapoda, Diogenidae), with notes on its behavior. *Pacific Science* 29, no. 3:259–266.

McMurtry, G. M., D. Epp, and D. M. Karl. 1983. Project Pele: Studies of the hydrology, chemistry and microbiology of geothermal systems on the submarine rift zones of the Hawaiian chain. *Sea Grant Quarterly* 5, no. 4:1–8.

Mead, G. 1972. Bramidae. *Dana Report,* no. 81:1–166.

Mochizuki, K., and M. Shirakihara. 1983. A new and rare apogonid species of the genus *Epigonus* from Japan. *Japanese Journal of Ichthyology* 30, no. 3:199–207.

Moffitt, R. B., and F. A. Parrish. 1992. An assessment of the exploitable biomass of *Heterocarpus laevigatus* in the main Hawaiian islands: 2. Observations from a submersible. *United States Fishery Bulletin* 90:476–482.

Moffitt, R. B., F. A. Parrish, and J. J. Polovina. 1989. Community structure, biomass and productivity of deepwater artificial reefs in Hawaii. *Bulletin of Marine Science* 44, no. 2:616–630.

Moore, J. G., and J. F. Campbell. 1987. Age of tilted reefs, Hawaii. *Journal of Geological Research* 92, no. B3:2641–2646.

Moore, J. G., and D. A. Clague. 1987. Coastal lava flows from Mauna Loa and Hualalai volcanos, Kona, Hawaii. *Volcanology* 49:752–764.

———. 1992. Volcano growth and evolution of the island of Hawaii. *Bulletin of the Geological Society of America* 104:1471–1484.

Moore, J. G., D. A. Clague, R. T. Holcomb, P. W. Lipman, W. R. Normark, and M. E. Torresan. 1989. Prodigious submarine landslides on the Hawaiian ridge. *Journal of Geophysical Research* 94, no. B12:17465–17484.

Moore, J. G., D. A. Clague, and W. R. Normark. 1982. Diverse basalt types from Loihi seamount, Hawaii. *Geology* 10:88–92.

Mortensen, T. 1928. *A monograph of the Echinoidea*. Vol. 1, *Cidaroidea*. Copenhagen: C. A. Reitzel. (551 pp.)

———. 1940. *A monograph of the Echinoidea*. Vol. 3, pt. 1, *Aulodonta*. Copenhagen: C. A. Reitzel. (370 pp.)

———. 1948. *A monograph of the Echinoidea*. Vol. 4, pt. 2, *Clypeasteroida*. Copenhagen: C. A. Reitzel. (223 pp.)

———. 1950. *A monograph of the Echinoidea*. Vol. 5, pt. 1, *Spatangoida*. Copenhagen: C. A. Reitzel. (217 pp.)

Musik, K. 1978. A bioluminescent gorgonian, *Lepidisis olapa*, new species (Coelenterata: Octocorallia) from Hawaii. *Bulletin of Marine Science* 28:735–741.

Myers, R. F. 1989. *Micronesian reef fishes*. Agana, Guam: Coral Graphics. (298 pp.)

Nakabo, T. 1982. Revision of the family Draconettidae. *Japanese Journal of Ichthyology* 28(4):355–367.

Nesis, K. N. 1982. *Cephalopods of the world.* Neptune City, N.J.: T.F.H. Publications. (242 pp.)

Newman, W. A. 1986. Origin of Hawaiian marine fauna: Dispersal and vicariance as indicated by barnacles and other organisms. In *Crustacean issues,* ed. R. H. Gore and K. L. Heck, 421–449. Rotterdam: A. A. Balkema.

Nishida, K. 1990. Phylogeny of the suborder Myliobatiodidae. *Memoirs of the Faculty of Fisheries, Hokkaido University* 37, nos. 1–2, serial no. 54:1–108.

Nishida, K., and K. Nakaya. 1990. Taxonomy of the genus *Dasyatis* (Elasmobranchii, Dasyatidae) from the North Pacific. *National Oceanic and Atmospheric Agency–National Marine Fisheries Service Technical Report* 90:327–346.

Novikov, N. P., L. S. Kodolov, and G. M. Gavrilov. 1980. Preliminary list of fishes of the Emperor underwater ridge. In *Fishes of the open ocean,* 32–35. Moscow: P. P. Shirshov Institute of Oceanology. (Translation of paper issued by the institute.)

Nutting, C. C. 1905. Hydroids of the Hawaiian Islands collected by the Steamer *Albatross* in 1902. In *Aquatic resources of the Hawaiian Islands,* ed. D. S. Jordan. *Bulletin of the United States Fish Commission for 1903* 23, pt. 3:933–959.

———. 1908. Descriptions of the Alcyonaria collected by the Steamer *Albatross* in the vicinity of the Hawaiian Islands in 1902. *Proceedings of the United States National Museum* 34, 1624:543–601.

Ohta, S. 1983. Photographic census of large-sized benthic organisms in the bathyal zone of Suruga Bay, central Japan. *Bulletin of the Ocean Research Institute, University of Tokyo* 15:1–244.

Okamoto, H., and B. Kanenaka. 1984. Preliminary report on the nearshore fishery resource assessment of the Northwestern Hawaiian Islands, 1977–1982. In *Proceedings of research investigations in the Northwestern Hawaiian Islands* (University of Hawai'i–Seagrant MR-84-01), 123–143. Honolulu.

Okamura, O. 1985. *Fishes of the Okinawa Trough and the adjacent waters.* Pt. 2. Tokyo: Japanese Fishery Research Conservation Association. (364 pp.)

Okamura, O., K. Amaoka, and F. Mitani. 1982. *Fishes of the Kyushu-Palau Ridge and Tosa Bay.* Tokyo: Japanese Fishery Research Conservation Association. (435 pp.)

Okamura, O., and T. Kitajima. 1984. *Fishes of the Okinawa Trough and the adjacent waters.* Pt. 1. Tokyo: Japanese Fishery Research Conservation Association. (414 pp.)

Okuda, R. K., D. Klein, R. B. Kinnel, M. Li, and P. J. Scheuer. 1982. Marine natural products: The past twenty years and beyond. *Pure and Applied Chemistry* 54, no. 10:1907–1914.

Opresko, D. M. 1974. A study of the classification of the Antipatharia (Coelenterata: Anthozoa) with description of eleven species. Ph.D. thesis, University of Miami. (373 pp.)

Parin, N. V. 1982. New species of the genus *Draconetta* and a key for the family Draconettidae (Osteichthyes). *Journal of Zoology* 61:544–563.

———. 1984. Three new species of *Physiculus* and other fishes (Moridae) from the submarine seamounts of the SE Pacific Ocean. *Journal of Ichthyology* 24:46–60.

———. 1989. Review of the genus *Rexea* (Gempylidae), with a description of three new species. *Journal of Ichthyology* 29, no. 2:86–105.

———. 1991. Fish fauna of the Nazca and Sala y Gomez submarine ridges, Indo-West Pacific region. *Bulletin of Marine Science* 49, no. 3:671–683.

Parin, N. V., and O. D. Borodulina. 1986. Preliminary review of the benthopelagic fish genus *Antigonia* Lowe (Zeiformes, Caproidae). *P. P. Shirshov Institute of Oceanology* (Moscow) 121:141–172. (Translation of paper issued by the institute.)

Parin, N. V., and A. N. Kotlyar. 1989. A new aulopodid species, *Hime microps,* from the eastern South Pacific, with comments on geographic variations of *H. japonica. Japanese Journal of Ichthyology* 35, no. 4:407–413.

Pasternak, F. A. 1977. Antipatharia. *Galathea Report* 14:157–164.

Paulin, C. D. 1989. Review of the morid genera *Gadella, Physiculus* and *Salilota* (Teleostei: Gadiformes) with descriptions of seven new species. *New Zealand Journal of Zoology* 16:93–133.

Pyle, R. 1991. Rare and unusual marines: The Hawaiian deep anthias *Holanthias fuscipinnis* (Jenkins). *Freshwater and Marine Aquarium Magazine* 14, no. 12:cover, 74, 76, 78.

Pyle, R., and E. H. Chave. 1994. First record of the chaetodontid genus *Prognathodes* from the Hawaiian Islands. *Pacific Science* 48, no. 1:90–93.

Ralston, S., R. M. Gooding, and G. M. Ludwig. 1986. An ecological survey and comparison of bottom fish resource assessments (submersible versus handline fishing) at Johnston Atoll. *Fisheries Bulletin* 84, no. 1:141–155.

Randall, J. E. 1963. Review of the hawkfishes (family Cirrhitidae). *Proceedings of the United States National Museum* 114, no. 3472:389–451.

———. 1975. A revision of the Indo-Pacific angelfish genus *Genicanthus*, with descriptions of three new species. *Bulletin of Marine Science* 25, no. 3:393–421.

———. 1979. A review of the serranid fish genus *Anthias* of the Hawaiian Islands with descriptions of two new species. *Contributions in Science, Natural History Museum, Los Angeles County*, no. 302:1–13.

———. 1980a. New records of fishes from the Hawaiian Islands. *Pacific Science* 34, no. 3:211–232.

———. 1980b. Revision of the fish genus *Plectranthias* (Serranidae: Anthiinae) with descriptions of 13 new species. *Micronesica* 16, no. 1:101–187.

———. 1981a. Examples of antitropical and antiequatorial distribution of Indo-West-Pacific fishes. *Pacific Science* 35, no. 3:197–209.

———. 1981b. *Luzonichthys earlei*, a new species of anthiine fish from the Hawaiian Islands. *Freshwater and Marine Aquarium Magazine* 4, no. 9:13–18.

———. 1985. *Guide to Hawaiian reef fishes.* Newtown Square, Pa.: Harrowood. (74 pp.)

———. 1996. *Shore fishes of Hawaiʻi.* Vida, Oreg.: Natural World Press. (216 pp.)

Randall, J. E., G. R. Allen, and R. C. Steene. 1990. *Fishes of the Great Barrier Reef and Coral Sea.* Honolulu: University of Hawaiʻi Press. (507 pp.)

Randall, J. E., and C. H. Chen. 1985. First record of the labrid fish *Bodianus cylindriatus* (Tanaka) from the Hawaiian Islands. *Pacific Science* 39, no. 3:291–293.

Randall, J. E., and P. C. Heemstra. 1991. Revision of Indo-Pacific groupers (Perciformes: Serranidae: Epinephelinae), with descriptions of five new species. *Indo-Pacific Fishes,* no. 20:1–332.

Randall, J. E., and W. Klausewitz. 1973. A review of the triggerfish genus *Melichthys* with descriptions of a new species from the Indian Ocean. *Senckenberg Biologie* 54, nos. 1–3:52–69.

Randall, J. E., and A. Kotthaus. 1977. *Suezichthys tripunctatus,* a new deep-dwelling Indo-Pacific labrid fish. *"Meteor" Forschung Ergebnisse,* ser. D(24):33–36.

Randall, J. E., P. S. Lobel, and E. H. Chave. 1985. Annotated checklist of the fishes of Johnston Island. *Pacific Science* 39, no. 1:24–80.

Randall, J. E., K. Matsuura, and A. Zama. 1978. A revision of the triggerfish genus *Xanthichthys,* with description of a new species. *Bulletin of Marine Science* 28, no. 4:688–706.

Randall, J. E., and J. E. McCosker. 1992. Revision of the fish genus *Luzonichthys* (Perciformes: Serranidae: Anthiinae) with descriptions of two new species. *Indo-Pacific Fishes,* no. 21:1–21.

Randall, J. E., and S. Ralston. 1984. A new species of serranid fish of the genus *Anthias* from the Hawaiian Islands and Johnston Island. *Pacific Science* 38, no. 3:220–227.

Randall, J. E., T. Shimizu, and T. Yamakawa. 1982. A revision of the holocentrid fish genus *Ostichthys* with descriptions of four new species and a related new genus. *Japanese Journal of Ichthyology* 29, no. 1:1–26.

Randall, J. E., and P. Struhsaker. 1976. Description of the male of the Hawaiian angelfish *Genicanthus personatus. Bulletin of Marine Science* 26, no. 3:414–416.

———. 1981. *Naso maculatus,* a new species of acanthurid fish from the Hawaiian Islands and Japan. *Copeia,* no. 3:553–558.

Randall, J. E., and S. N. Swerdloff. 1973. A review of the damselfish genus *Chromis* from the Hawaiian Islands, with descriptions of three new species. *Pacific Science* 27, no. 4:327–349.

Randall, J. E., and L. Taylor. 1988. Review of the Indo-Pacific fishes of the serranid genus *Liopropoma* with descriptions of seven new species. *Indo-Pacific Fishes,* no. 16:1–47.

Rao, M. V., and W. A. Newman. 1972. Thoracic Cirripedia from guyots of the Mid-Pacific Mountains. *Transactions of the San Diego Society of Natural History* 17:69–94.

Rathbun, M. J. 1906. The Brachyura and Macrura of the Hawaiian Islands. In *Aquatic resources of the Hawaiian Islands*, ed. D. S. Jordan. *Bulletin of the United States Fish Commission for 1903* 23, pt. 3:827–930.

Rehder, H. A., and H. S. Ladd. 1973. Deep and shallow-water mollusks from the central Pacific. *Tohoku University Science Reports* 6:37–51.

Reid, S. B., J. Hirota, R. E. Young, and L. E. Hallacher. 1991. Mesopelagic-boundary community in Hawaii: Micronekton at the interface between neritic and oceanic ecosystems. *Marine Biology* 109:427–440.

Robilliard, G. A., and P. K. Dayton. 1972. A new species of platyctenean ctenophore, *Lyrocteis flavopallidus* sp. nov., from McMurdo Sound, Antarctica. *Canadian Journal of Zoology* 50:47–52.

Robins, C. H., and C. R. Robins. 1989. Family Synaphobranchidae. In *Fishes of the western North Atlantic,* ed. D. M. Cohen et al. *Memoirs of the Sears Foundation for Marine Research* (Yale University) 1, pt. 9:207–253.

Roper, C. F. E., M. J. Sweeney, and C. E. Nauer. 1984. Cephalopods of the world (F.A.O. Species Catalogue). *F.A.O. Fisheries Synopsis*, no. 125, vol. 3, pt. 1, pp. 1–277.

Roux, M. 1994. The CALSUB cruise on the bathyal slopes off New Caledonia. In *Resultats des campagnes MUSORSTOM,* vol. 12, ed. A. Crosnier. *Memoires de la Musee Nationale Histoire Naturelle* (Paris) 161:9–47.

Rowe, G. T., and J. E. Doty. 1977. The shallow-water holothurians of Guam. *Micronesica* 13, no. 2:217–250.

Russell, B. 1985. Revision of the Indo-Pacific labrid fish genus *Suezichthys,* with descriptions of four new species. *Indo-Pacific Fishes*, no. 2:1–21.

Russo, R. 1994. *Hawaiian reefs: A natural history guide.* San Leandro, Calif.: Wavecrest. (174 pp.)

Sager, W. W., and M. S. Pringle. 1987. Paleomagnetic constraints on the origin and evolution of the Musician and southern Hawaiian Seamounts, central Pacific Ocean. In *Seamounts, islands, and atolls,* ed. B. H. Keating, P. Fryer, R. Batiza, and G. W. Boehlert, 133–162. American Geophysical Union Monograph 43. Washington, D.C.

Sakai, K. 1987. Biogeographical records of five species of the family Lithodidae from the abyssal valley off Gamoda-Misaki, Tokushima, Japan. *Researches on Crustacea* 16:19–24.

Sakai, T. 1978. Decapod crustacea from the Emperor seamount chain. *Researches on Crustacea* 8, suppl.:1–39.

Sazonov, Y. I., and T. Iwamoto. 1992. Grenadiers (Pisces, Gadiformes) of the Nazca and Sala y Gomez Ridges, Southeastern Pacific. *Proceedings of the California Academy of Sciences* 48, no. 2:27–95.

Schultze, F. E. 1886. Uber den Bau und das System der Hexanelliden. *Abhandlungen der Koniglichen Akademie der Wissenschaften zu Berlin* (Physikalisch-Mathematische Classe), 1–97.

———. 1887. Report on the Hexactinellida collected by the H.M.S. *Challenger* during the years 1873–1876. In *Report of the scientific results of the voyage of the H.M.S. Challenger during the years 1873–1876* (Zoology section 5), ed. C. W. Thompson and J. Murray, 21:1–513. London: Eyre & Spottiswoode.

———. 1899. Amerikanische Hexactinellidea nach dem materiale der *Albatross* Expedition. In *Preussische Akadamie Wissenschafte,* 1–261. Jena: Fischer.

Sedwick, P. N., G. M. McMurtry, and J. D. Macdougall. 1992. Chemistry of hydrothermal solutions from Pele's vents, Loihi Seamount. *Geochimica Cosmochimica Acta* 56:3643–3667.

Severns, M., and P. F. Fiene-Severns. 1993. *Molokini Island, Hawaii's premier marine preserve.* Honolulu: Pacific Islands Publications. (144 pp.)

Shcherbachev, Y. N., E. I. Kukuev, and V. I. Shilbanov. 1985. Composition of the benthic and demersal ichthyocenoses of the submarine mountains in the southern part of the North Atlantic range. *Journal of Ichthyology* 25:110–125.

Sladen, W. P. 1889. Report on the Asteroidea collected by the H.M.S. *Challenger.* In *Report of the scientific results of the voyage of the H.M.S. Challenger during the years 1873–1876* (Zoology section 5), ed. C. W. Thompson and J. Murray, 30, no. 1:1–222. London: Eyre & Spottiswoode.

Small, G. J. 1981. A review of the bathyal fish genus *Antimora. Proceedings of the California Academy of Sciences* 42, no. 13:341–348.

Smith, D. G. 1989. Family Congridae. In *Fishes of the western North Atlantic,* ed. D. M. Cohen et al. *Memoirs of the Sears Foundation for Marine Research* (Yale University) 1, pt. 9:460–567.

Smith, M. M., and C. Heemstra. 1986. *Smith's sea fishes.* Berlin: Springer. (1,047 pp.)

Soule, J. D., D. T. Soule, and H. W. Chaney. 1986. Phyla Entoprocta and Bryozoa (Ectoprocta). In *Reef and shore fauna of Hawaii,* vol. 2, ed. D. M. Devaney and L. G. Eldredge, 83–166. Bernice P. Bishop Museum Special Publication 64. Honolulu.

Springer, V. G. 1982. Pacific Plate biogeography with special reference to shore-fishes. *Smithsonian Contributions in Zoology,* no. 367:1–182.

Starnes, W. C. 1988. Revision, phylogeny and biogeographic comments on the circumtropical marine percoid fish family Priacanthidae. *Bulletin of Marine Science* 43, no. 2:117–203.

Stearns, H. T. 1966. *Geology of the state of Hawaii.* Palo Alto, Calif.: Pacific Books. (262 pp.)

———. 1974. Submerged shorelines and shelves in the Hawaiian Islands and a review of some of the eustatic emerged shorelines. *Geological Society of America Bulletin* 85:795–804.

Strasburg, D. W., E. C. Jones, and R. T. B. Iverson. 1968. Use of a small submarine for biological and oceanographic research. *Journal du Conseil International pour l'Exploration de la Mer* 31, no. 3:410–426.

Struhsaker, P. 1973. A contribution to the systematics and ecology of Hawaiian bathyal fishes. Ph.D. thesis, University of Hawai'i. (482 pp.)

Sulak, K. J. 1977. The systematics and biology of *Bathypterois* (Pisces, Chlorophthalmidae) with a revised classification of benthic myctophiform fishes. *Galathea Report* 14:49–108.

Szabo, K. R., J. G. Moore, and K. R. Simmons. 1991. Crustal subsidence rate off Hawaii determined from $^{234}U/^{238}U$ ages of drowned coral reefs. *Geology* 19:171–174.

Tabachnick, K. R. 1981. *Benthos of the submarine mountains of Marcus-Necker and adjacent Pacific regions.* Moscow: P. P. Shirshov Institute of Oceanology. (149 pp.)

———. 1988. Hexactinellid sponges from the mountains of west Pacific. In *Structural and functional researches of the marine benthos,* ed. A. P. Kushetzov and M. N. Sokolova, 49–64. Moscow: P. P. Shirshov Institute of Oceanology.

———. 1991. Adaptation of hexactanellid sponges to deep-sea life. In *Fossil and recent sponges,* ed. J. Reitner and H. Krupp, 378–386. Berlin: Springer.

Takeda, M. 1974. On three species of the Lithodidae (Crustacea, Anomura) from the central Pacific. *Bulletin of the National Science Museum,* Tokyo 17, no. 3:205–214.

———. 1977. Two interesting crabs from Hawaii. *Pacific Science* 3, no. 1:31–38.

Takeda, M., and T. Okutani. 1983. *Crustaceans and mollusks trawled off Suriname and French Guiana.* Tokyo: Kaiyo Suisan Shigen Kaihatsu Senta. (354 pp.)

Théel, H. 1882. Report on the Holothuroidea dredged by the H.M.S. *Challenger* during the years 1873–1876. In *Report of the scientific results of the voyage of the H.M.S. Challenger during the years 1873–1876* (Zoology section 5), ed. C. W. Thompson and J. Murray, 4, no. 3:1–176. London: Eyre & Spottiswoode.

———. 1886. Report on the Holothuroidea dredged by the H.M.S. *Challenger* during the years 1873–1876. In *Report of the scientific results of the voyage of the H.M.S. Challenger during the years 1873–1876* (Zoology section 5), ed. C. W. Thompson and J. Murray, 14, no. 2:1–290. London: Eyre & Spottiswoode.

Tinker, S. W. 1978. *Fishes of Hawaii.* Honolulu: Hawaiian Service. (532 pp.)

Titgen, R. H. 1987. New decapod records from the Hawaiian Islands. *Pacific Science* 41, nos. 1–4:141–147.

Tunnicliffe, V. 1992. Hydrothermal-vent communities of the deep-sea. *American Scientist* 80:336–349.

Uchida, R. N., and D. T. Tagami. 1984. Groundfish fisheries and research in the vicinity of seamounts in the north Pacific Ocean. *Marine Fisheries Review* 46, no. 2:1–17.

Vaughan, T. W. 1907. Recent Madreporaria of the Hawaiian Islands and Laysan. *Bulletin of the United States National Museum* 59:1–84.

Waples, R. S., and J. E. Randall. 1988. A revision of the Hawaiian lizardfishes of the genus *Synodus,* with descriptions of four new species. *Pacific Science* 42, nos. 3–4:178–213.

Westneat, M. W. 1993. Phylogenetic relationships of the tribe Cheilinini (Labridae: Perciformes). *Bulletin of Marine Science* 52, no. 1:351–394.

Wicksten, M. 1985. Carrying behavior in the family Homolidae (Decapoda: Brachyura). *Journal of Crustacean Biology* 5, no. 3:476–479.

Williams, A. B. 1982. Revision of the genus *Latreillia* Roux (Brachyura, Homoloidea). *Quaderni del Laboratorio di Tecnologica Della Pesca* 3, nos. 2–5:227–255.

Williams, A. B., and F. C. Dobbs. 1995. A new genus and species of caridean shrimp (Crustacea: Decapoda: Bresiliidae) from hydrothermal vents on Loihi Seamount, Hawaii. *Proceedings of the Biological Society of Washington* 108, no. 2:228–237.

Wilson, R. R., Jr., and R. S. Kaufmann. 1987. Seamount biogeography. In *Seamounts, islands, and atolls*, ed. B. H. Keating, P. Fryer, R. Batiza, and G. W. Boehlert, 355–377. American Geophysical Union Monograph 43. Washington, D.C.

Wilson, R. R., Jr., K. L. Smith, and R. H. Rosenblatt. 1985. Megafauna associated with bathyal seamounts in the central North Pacific Ocean. *Deep-Sea Research* 32, no. 10:1243–1254.

Wiltshire, J. 1982. The potential of the Puna submarine canyon for slurry disposal of manganese nodule tailings. *Oceans* (September), 1069–1073.

Wright, E. P., and T. Studer. 1889. Report on the Alcyonaria collected by H.M.S. *Challenger* during the years 1873–1876. In *Report of the scientific results of the voyage of the H.M.S. Challenger during the years 1873–1876* (Zoology section 5), ed. C. W. Thompson and J. Murray, 31:1–314. London: Eyre & Spottiswoode.

Young, C. M., P. A. Tyler, J. L. Cameron, and S. G. Rumrill. 1992. Seasonal breeding aggregations in low-density populations of the bathyal echinoid *Stylocidaris lineata*. *Marine Biology* 113:603–612.

Young, R. E. 1995. Aspects of the natural history of pelagic cephalopods of the Hawaiian mesopelagic-boundary region. *Pacific Science* 49, no. 2:143–155.

General Index

Page numbers in **boldface** denote illustrations.

ash (tuff), **14,** 16, 20
attachment of organisms to bottom, 30, 35–38, 41–42
bacterial mat, **16,** 27
basalt: outcrop, 19; pinnacle, 17–18
bioluminescence, 35, 37, 42, 70
bottom topography, 26–27
boulder, **18–19**
chimney, 16
commercial value, of deep-sea animals, 33, 35, 39, 41–42, 54, 65, 70
coral beds, Makapuʻu, 20
coral reef: communities, 25; fossil, 10–**12,** 13, 15–16, **18–19,** 20–22, **53;** modern, 10, 15, 18, 20
coralline alga, **25,** 64
Cross Seamount, **3,** 18
currents, 13, 23–26
defense, 30, 41, 48, 53–54, 59, 62, 69–70
dike, 16, 18
distribution of animals, 23–24
diurnal and nocturnal patterns, 62, 65, 67, 70, 73–74
earthquake 16
East Pacific Rise, 9, 10, 18
feeding, 28, 30, 37–38, 42, 44, 47–48, 50, 53–54, 56, 58–59, 62, 67–70, 74–75
ferromanganese crust, **14, 18,** 38

fissure, 16
food supply, 25–26
group size, animal, 62, 65, 67–70
habitat, 30, 36, 39–44, 47–48, 50, 53–54, 56–59, 62, 65, 68–70, 72–73
hardpan, **14,** 16, 18, **53**
Hawaiʻi (Island), **3,** 9, 15–16
"hot spot" volcano, 9, 15, 18
hydrothermal vent, **16–17;** biota, **27;** minerals, 16
Johnston Atoll, 10, 20–**21,** 22
kīpuka, **27**
Lānaʻi, **3,** 9, 18–22
larval transport, 23–24
lava: magma, 9, 11, 16; pillow, 11–**12,** 16–17, **34;** types of, 11–**12,** 15–16
lava bomb, 20
light intensity, 25–26
limestone: bench (terrace), **12**–13, **19–20;** crust, 13–**14;** eroded, 13, 15, **20–22**
locomotion, 43–44, 47–48, 50, 53–54, 70, 73–75
Lōʻihi submarine volcano, **3, 7,** 9, 13, 16–17
Makapuʻu coral beds, 20
Maui, **3,** 9, 19
Molokaʻi, **3,** 9, 18–19
Northwestern Hawaiian Islands, 9–10, 20
Oʻahu, **3,** 9, 19
origin of animals, 23–24
Peleʻs Vents, **16–17**

Penguin Bank, 18
pillow lava, 11–**12,** 16–**17, 34**
plate tectonics, 9
reef communities, **25**
response to external stimuli, 44, 48, 50, 53–54, 56–57, 62, 67, 69–70, 72, 74
rift features, 15–16
sand, movement of, 26, 30
sediment: movement of, 13, 19, 22; types of, **12**–13, 15, **18–20,** 22
sediment plain, 13, 15–16, 19–20, 22
sex and reproduction, 47–48, 67–68
skeletal structure, 28, 30–31, 33, 35, 42, 53, 58
slumping, of seamount 16, 18
subsidence, 10, 20
talus, 13–**14,** 15–16, 18, **34**
temperature, **16,** 26
volcanic cone, 16, 20
volcanic rift, 15–16
volcano: Hawaiian, 9, 10, 15; hot spot, 9, 15, 18; submarine (Lōʻihi), **3, 7,** 9, 13, 16–17
water: chemistry, 26–27; temperature, **16,** 26

Systematic Index

ANNELIDA (segmented worms), 29, 38
ARTHROPODA
  Crustacea
    barnacle, 40, 42, 51, 53
    bresiliid shrimp, 24, 27
    crab, 29, 42, 53–54
    galatheid crab, 29, 32, 54
    lithodid crab, 26, 53
    lobster, 54
    pagurid (hermit crab), 32, 38, 55
    prawn, 54–55, 64
    shrimp, 26, 29, 55
BRACHIOPODA, 19
CHORDATA
  Chondrichthys (sharks, rays, chimeras)
    Carcharhinidae (reef sharks and others), 58–59
    Chimaeridae, 59, 61
    Echinorhinidae (Cook's shark and others), 58–59
    Hexanchidae (six-gilled sharks), 58–59
    Hexatrygonidae (longnosed rays), 59, 61
    Plesiobatidae (round rays), 59–60
    Rhinochimaeridae (longnosed chimeras), 59, 61
    Squalidae (dogfish), 58–59
    Torpedinidae (electric ray), 59, 61

Osteichthys (bony fishes)
  Acanthuridae (surgeonfish, tang), 25, 66
  Acropomatidae (lanternbelly), 70
  Ateliopidae (jellynose eel), 72
  Berycidae (alfonsin), 70–71
  Bothidae (flatfish), 70–71
  Caproidae (boarfish), 69
  Carangidae (jack, mackerel, rainbow runner), 62–63
  Chaetodontidae (butterflyfish), 21, 25, 54, 67–68
  Chaunacidae (sea toad), VII, 75
  Chloropthalmidae (greeneye), 73
  Congridae (conger eel), 62, 64
  Gempylidae (snake mackerel), 70–71
  Halosauridae, 72
  Holocentridae (soldierfish, squirrelfish), 36, 67–68
  Ipnopidae (spiderfish, tripodfish), 72–73, 74
  Labridae (wrasse), 25
  Lophiidae (goosefish), 74–75
  Lutjanidae (snapper), 65–66, 67–68, 69
  Macrouridae (grenadier, rattail), 40, 74
  Moridae (cod), 19, 74
  Mullidae (goatfish), 54, 67
  Muraenidae (moray eel), 11
  Nettastomatidae (duckbill eel), 72
  Ophichthidae (snake eel), 62, 64
  Ophidiidae, 72
  Owstoniidae (bandfish), 69
  Percophidae (flathead), 69
  Polymixiidae (beardfish), 74
  Pomacentridae (damselfish), 25, 54, 67
  Priacanthidae (bigeye), 68
  Scorpaenidae (scorpionfish), 13, 67, 69
  Serranidae (basslet, grouper), 67–68
  Synaphobranchidae (cutthroat and arrowtooth eels), 62, 64
  Tetradontidae (pufferfish), 70
  Triacanthodidae (spikefish), 70
  Triglidae (rakefish), 19, 70–71
  Zeidae (john dory), 70–71
CNIDARIA (corals)
  Actiniaria (anemones), 37–40
    black and white, 26, 39
    flytrap anemone, 26, 38–39, 52
  Alcyonacea, 36–37
    neptheid, 22, 36
    toadstool corals, VII, 37, 56
  Antipatharia (black corals, wire corals), 19, 39, 40–41, 42
  Ceriantharia, 38
  Corallimorpharia, 37–38

Gorgonacea (sea fans, sea whips), **26**, **31**–**32**
  Acanthogorgidae, **31**, 33
  Briareidae (*Paragorgia*), **27**, **35**–**36**
  Chrysogorgiidae, **17**, **27**, **33**–**34**, 35
  Corallidae (pink coral), **35**, **37**, 42
  Ellisellidae, **27**, **45**
  Isididae (bamboo corals), **20**, **34**–**35**, 41
  Paramuricidae, **33**, **50**, **52**
  Primnoidae, **18**, **20**, **28**, **31**–**32**, **39**, **42**–**43**, **64**
Hydroida, 36, 40, 42
Milliporina, 36
Pennatulacea (sea pen, rock pen), **37**–**38**, 43
Scleractinia (stony corals)
  branched, **35**, 42, 67
  reef building, **10**, **19**, **25**
  solitary, **11**, **22**, 40
Stolonifera, 37
Stylasterina, **36**, 48
Zoanthinaria (parasitic corals), 30, 40, 41–42
gold coral, **41**, **64**
CTENOPHORA (comb jellyfish), **37**–**38**
ECHINODERMATA
  Asteroida (sea stars), **43**–**45**
    Brisingidae, **19**, **44**–**45**
    Goniasteridae, **13**, **43**–**44**, **45**, **64**
  Crinoidea (crinoids and sea lilies), 50
    stalked, **19**, **26**, **50**, **52**
    unstalked, 42, **50**, **51**–**52**, **69**
  Echinoida (sea urchins), **18**, **25**, **44**, **46**, **47**–**48**
    Cidaridae, **15**, 40, 42, **46**–**47**
    Diadematidae (long-spined), **43**–**44**, **48**
  Holothuroida (sea cucumbers), **48**–**49**
  Ophiuroida (brittle stars), **17**–**18**, **26**–**27**, 38, **45**, **47**–**48**, **50**, **52**
    Gorgonocephalidae (basket star), **50**–**51**
MOLLUSCA
  bivalve, **45**, 56
  gastropod, 56
  nudibranch, **56**
  cephalopod (octopus), **56**–**57**
POGONOPHORA, 27
PORIFERA (sponges)
  Demospongia
    plate sponge, **29**
    encrusting sponges, 28, 36
  Hexactanellida (glass sponges), **28**–**30**, 56
    Caulophacidae (stalked), **19**
    Dactylocalicidae (vase), **29**
    Euplectellidae, 28–30
      bubble, **28**
      feather, **17**, **28**
      finger, **18**, **28**
      porous, **30**
      stalked, **28**–**29**
    Farreidae, **26**, **28**
    Hyalonematidae (rope), **30**, 53
    Pheronematidae, 30
      satin, **30**
      scoop, **30**

About the Authors

E. H. CHAVE is the biological researcher at the Hawai'i Undersea Research Laboratory, University of Hawai'i. She received her M.A. and B.A. from the University of California, Berkeley, transferred to the University of Hawai'i from Stanford University in 1967, and received her Ph.D. there in 1971. She has held research and teaching positions at the University of Hawai'i since 1972 with the Waikiki Aquarium, Marine Programs, and Sea Grant. In 1981 she joined the Hawai'i Undersea Research Laboratory (HURL) as science director and has since held various positions there.

Her research for the first decade focused on Pacific reef fishes and coral reef communities. For the last fifteen years she has been collecting, disseminating, and publishing various kinds of data collected by the HURL submersibles. She has authored over fifty scientific papers, mainly on fishes and foraminiferans. She has coauthored two books, including *Hawaiian Reef Animals* (University of Hawai'i Press). Her most recent work consisted of developing the databases on which the material for this book is founded.

ALEXANDER MALAHOFF holds the position of professor of geological oceanography in the Department of Oceanography, University of Hawai'i. He is also the director of the National Oceanic and Atmospheric Agency's National Undersea Research Center at the University of Hawai'i (HURL). He received his B.S. in geology from the University of New Zealand in 1960, his M.S. in geology from Victoria University of Wellington in 1962, and his Ph.D. in geophysics from the University of Hawai'i in 1965. He has held research positions with the Department of Scientific and Industrial Research, New Zealand, and the University of Wisconsin and has served as program director for the Marine Geology and Geophysics Program, Office of Naval Research. In 1976 he became chief scientist for the National Ocean Survey, National Oceanic and Atmospheric Administration, before returning to the University of Hawai'i in 1984.

His research over the past three decades has been focused on the geology and geophysics of the ocean floor and of volcanos and volcanic islands, using ships, airplanes, and submarines in order to conduct these studies. He has been actively involved in the design and reconstruction of vehicles, such as submersibles and research ships, and the development of specialized instrumentation, such as remotely controlled vehicles and ocean bottom observatories. He is the author of over seventy scientific papers. Much of his recent work has centered around studies of hydrothermal vent processes on submarine volcanos and cobalt-rich ferromanganese crusts around the Hawaiian Islands.